T0296096

OUTLINE OF WIRELESS

WIRELESS TELEGRAPHY AND TELEPHONY

AN OUTLINE FOR ELECTRICAL ENGINEERS AND OTHERS

BY

L. B. TURNER, M.A., M.I.E.E.

FELLOW OF KING'S COLLEGE, CAMBRIDGE;
RECENTLY OF THE SIGNALS EXPERIMENTAL ESTABLISHMENT
WOOLWICH (ROYAL ENGINEERS),
AND OF THE ENGINEER-IN-CHIEF'S OFFICE, POST OFFICE;
MEMBER OF THE IMPERIAL WIRELESS TELEGRAPHY COMMITTEE 1919–20.

CAMBRIDGE
AT THE UNIVERSITY PRESS
1921

CAMBRIDGE UNIVERSITY PRESS
Cambridge, New York, Melbourne, Madrid, Cape Town,
Singapore, São Paulo, Delhi, Mexico City

Cambridge University Press
The Edinburgh Building, Cambridge CB2 8RU, UK

Published in the United States of America by Cambridge University Press, New York

www.cambridge.org
Information on this title: www.cambridge.org/9781107629561

© Cambridge University Press 1921

First published 1921
First paperback edition 2013

A catalogue record for this publication is available from the British Library

ISBN 978-1-107-62956-1 Paperback

PREFACE

THE rapid development, scientific, commercial and military, of wireless telegraphy during the last decade has called forth numerous textbooks of very varying quality, and the second and third of three important scientific periodicals (German, American and English) devoted solely to the theory and practice of wireless. The Institution of Electrical Engineers, too, has formed a Wireless Section, and publishes in the Journal the papers read before it. This literature more or less adequately meets the requirements of the wireless operator on the one hand, and of the specialist wireless engineer on the other; but it does not provide for the large class of interested persons occupying an intermediate position, such as that of the electrical engineer who has never studied this particular branch of his general subject.

In this book an attempt is made to fill the gap. I have endeavoured to keep before me, especially in the earlier chapters, the standpoint of the engineer acquainted with the mathematical and electrical principles involved, but ignorant of this highly specialised branch of electrical engineering. Methods and principles have been treated as more significant than details of practice, although as much of the latter has been included as will enable the reader to envisage the types of apparatus and the orders of magnitude involved in the processes under discussion. Throughout I have striven towards brevity, and have dealt with each topic in a typical rather than an exhaustive manner; but I have not hesitated to introduce new matter, or to discuss old questions from a fresh standpoint, wherever this has seemed to facilitate the end in view. The book is an outline only of the framework of a great and growing subject; but it is hoped that a careful perusal will enable the intelligent reader to appreciate the problems which are presenting themselves, and to read, if he so desire, with understanding any of the discussions of these problems appearing in the technical press. In this sense, and in no other, can the work be regarded as complete.

If the book is successful in this its main aim, it may also prove useful as the groundwork of a course of instruction for engineering students in universities and technical colleges, where a textbook of similar scope and character seems required. Wireless Telegraphy is a difficult subject to compress into an engineering curriculum; but a large demand for trained wireless engineers is certain to arise in the near future, and wherever electrical specialisation is undertaken, post-graduate or other advanced courses in wireless will have to be provided. Such courses, to be efficient, must be accompanied by laboratory work. Fortunately, the obvious trend of modern development—towards the use of the thermionic tube at every stage—makes for the easy provision of effective laboratory experiments closely representing practical working conditions; and the possibility of precise measurements with such apparatus gives great educational value to such experimental work. The thermionic methods of wireless, too, offer excellent subjects for fruitful researches on a small scale. These methods have claimed a large share of the present volume. It is hoped that the manner in which they are here approached may help to evoke a proper enthusiasm for the triumphs of the last few years and the greater possibilities which lie ahead.

To students wishing to read outside this volume, the following works are recommended as particularly helpful in their several ways. *The Principles underlying Radio Communication (Radio Pamphlet, no. 40)*, prepared by the Bureau of Standards, Washington, for use in the training of military "radio electricians," is an elementary and non-mathematical book of quite exceptional merit, which can be read with profit by anyone from operator upwards; Professor J. Zenneck's *Lehrbuch der drahtlosen Telegraphie* (of which there is an English translation) is an admirably careful and lucid comprehensive textbook; and Dr W. H. Eccles' handbook, *Wireless Telegraphy and Telephony*, is a mine of information presented with scholarly precision. The *Proceedings of the Institute of Radio Engineers* (New York) and the *Radio Review* (edited by Professor G. W. O. Howe) are excellent periodicals which will keep the reader in touch with current work. A valuable feature of the latter is its monthly review of wireless articles and patents.

I take this opportunity of expressing my admiration of the achievements of General G. Ferrié, C.M.G. and his brilliant staff at the French "Télégraphie Militaire" in the development and application of the high-vacuum three-electrode thermionic tube; and my personal indebtedness, as a humble fellow-worker during the war, to the illuminating monographs issued from that institution.

To Mr E. B. Moullin, M.A., I tender my thanks for his kindness in reading the proofs of this book.

L. B. T.

ENGINEERING LABORATORY,
CAMBRIDGE.
November 1920.

TABLE OF CONTENTS

CHAPTER V,

THE DETECTION OF HIGH-FREQUENCY CURRENTS

CHAPTER VI

THE THERMIONIC TUBE

CHAPTER VII

THE TRIODE AS AMPLIFIER

CHAPTER VIII

THE TRIODE AS RECTIFIER

CHAPTER IX

THE TRIODE AS OSCILLATION GENERATOR

CHAPTER X

RETROACTIVE AMPLIFICATION AND RECTIFICATION

CHAPTER XI

WIRELESS TELEPHONY

CHAPTER XII

MISCELLANEOUS

ERRATUM

p. 85. In Fig. 56 for *Amperes* read *Milliamperes*.

LIST OF PLATES

In accordance with the recommendations of the International Electrotechnical Commission, the following signs for name of units are used:

A for ampere	F for farad
V ,, volt	H ,, henry
Ω ,, ohm	m ,, metre
W ,, watt	

Prefix μ $\equiv$ micro-, e.g. $1\mu\mu F \equiv 1$ micromicrofarad $\equiv 10^{-12}$ farad

,,   m $\equiv$ milli-,   ,, $1mA \equiv 1$ milliampere $\equiv 10^{-3}$ ampere

,,   k $\equiv$ kilo-,   ,, $1kW \equiv 1$ kilowatt $\equiv 10^{3}$ watts

,,   M $\equiv$ mega-,   ,, $1\ M\Omega \equiv 1$ megohm $\equiv 10^{6}$ ohms

Throughout the book letters standing for quantities are in italic type. Also, in the case of an article such as a condenser or an inductance coil, an italic letter is used to designate both the article and its size. Thus:

C_1 is a condenser of capacity C_1

L_2 is a coil of inductance L_2

Plate I. T-antenna at Colombo (p. 9).
(The Wireless Press, Limited.)

PLATE II. COLLAPSIBLE LOOP ANTENNA FOR PORTABLE TRANSMITTER (p. 9).

CHAPTER I

INTRODUCTION

WIRELESS TELEGRAPHY to-day is a large and rapidly expanding subject. Mathematician, physicist and engineer have contributed to its growth, and it has already become on the one side a highly developed exact science, and on the other a branch of engineering of great utilitarian value. This book is concerned directly only with the latter aspect of the subject, and any mathematical or physical theory included is introduced with the specific purpose of explaining practical applications. As an engineering subject, wireless telegraphy is of great importance from several points of view. Progress has been, and is being, made with remarkable rapidity at the hands of a multitude of workers; partly perhaps owing to the obvious fascination of the subject as a hobby; partly, it is probable, on account of the commercial and humanitarian importance of telegraphic connection with ships at sea; and partly, without doubt, owing to the military significance of wireless by land and sea and air. This large and enlarging position occupied by wireless amongst the amenities of mechanical civilisation involves the need of numerous constructing, operating and controlling staffs; so that wireless as a branch of electrical engineering is likely to enrol many men, and women, of high calibre and education.

In taking a preliminary view of wireless telegraphy as a whole, we may note first certain features which assimilate it to, or which differentiate it from, other branches of physics or electrical engineering. The essentials of a complete wireless installation are (i) a collection or box of instruments (often of very simple construction) in which alternating currents of very high frequency are produced from some local source of electric energy; (ii) an antenna, or electric circuit of such a geometrical form that high-frequency electric currents in it are accompanied by a marked radiation of energy into surrounding space; and elsewhere on or near the surface of the earth, (iii) another antenna, with (iv) another box of instruments in which alternating currents are produced by the tiny fraction of the radiated power which reaches them from the sending station.

Electromagnetic radiation, by which energy is transferred from the sending to the receiving antenna, is a phenomenon familiar to all who have eyes to see or a skin to feel the Sun's rays. It is, we believe, by electromagnetic radiation that light reaches us; and the same mechanism of the transference of energy with the same mathematical analysis suffices to describe radiation in the case of visible light, where the frequency is about 5×10^{14} periods per second, as in the case of wireless telegraphy, where the frequency commonly lies between 5×10^6 periods per second and one-hundredth of that number.

There are these contrasts to be noted, however. Firstly, in the case of light the frequency is so large, and therefore the wave-length· and the size of the radiating oscillator so small, that physicists have as yet been able to do little in the way of arbitrarily constructing and disposing their luminous oscillators, but must take them as they find them in the atom; whereas the wireless engineer builds his own radiator, his antenna, long or short, high or low, of this shape or that; for it is, as it were, large enough to give room for his fingers. Secondly, in wireless we are not so much concerned with radiation through free space as along the earth's surface. Even the aeroplane cannot get far enough from the earth to be regarded as unaffected by it. There is, moreover, the further complication of another conducting surface, the ionised upper atmosphere. So that radiation in wireless telegraphy is not through free space; or even over a plane conducting surface bounding free space, though this is sometimes a convenient approximation to the actual conditions; it occurs between an uneven heterogeneous spheroidal solid and liquid body, and the even less well-defined gaseous conducting layer. Consequently the exact mechanism of this radiation, whether at the antenna or far away between the antennae, is imperfectly understood and hard to ascertain.

The processes occurring within the boxes of instruments associated with the antennae can be analysed with greater precision and detail. They are those encountered in ordinary alternating current theory, with only such quantitative differences as follow from the much higher frequencies of the currents to be handled. The central station engineer who has studied alternating currents, including transient phenomena, is largely at home with the theory of wireless circuits. But as the admittance of a condenser and the impedance of an inductance are proportional to the frequency, with the vastly greater frequencies of wireless, tiny capacities and tiny inductances become important which would be utterly ignored in alternating current power circuits, or even in telephone circuits.

It is well to get an idea of actual values. In wireless, particularly of course in short-wave work, a capacity of (say) 1 micromicrofarad may be very perceptible. This is the capacity in air between two parallel sixpenny pieces spaced about twice their total thickness apart $(\frac{3}{32}'')$, or between the earth and a distant sphere of about 1 cm radius. At 3×10^6 periods per second (which corresponds to a wavelength of 100 metres), an inductance of 1 microhenry might be of equal importance; and this would be provided by two or three close turns of wire round an ordinary glass tumbler. Now the 50-cycle engineer does not worry about a micro-microfarad, for three thousand million of his volts would be needed to drive an ampere through it; nor does he appreciate a microhenry, for three thousand of his amperes would produce a P.D. of only 1 volt across it. The efficient wireless experimenter, however, must be constantly alive to the effects of such small capacities and inductances. He develops, for example, a habit of mind which classifies the points of a circuit as sacred, and profane or earthy, from the high-frequency aspect. The sacred point is one at which high-frequency potentials are developed, and no liberties must be taken there; the profane or earthy point is one at which no high-frequency potentials should be developed, and if any necessarily earthy instrument, such as a pair of headgear telephone receivers or a bulky battery, is to be inserted in the circuit it should be at this point. Fig. 1 illustrates this in the very simple case of a receiving circuit comprising the antenna A, the rectifier R and the telephone T. Capacity between telephone and earth would be without effect in the Right arrangement, but in the Wrong would shunt the rectifier and distune the antenna.

Regarded as a system of power transmission, wireless telegraphy occupies a peculiar place in the extreme smallness of the fraction of the transmitted power which reaches the receiver; and the close juxtaposition in a wireless station of the great and the minute, is, perhaps, responsible for some of the fascination, and certainly for some of the difficulties, experienced by its devotees. In the ordinary electric power line, energy may be poured into one end at an enormous rate, but at the other end the power is on a correspondingly large scale. In submarine telegraphy moderate power

is transmitted and very small power received; and in ordinary
telephony, the power transmitted is only a small fraction of a watt,

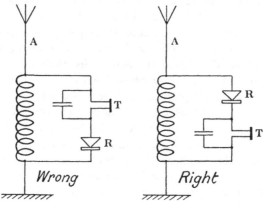

Fig. 1. The importance of small stray capacities.

and, over a line with the conventional commercial limit of attenuation ($\beta l \approx 5$), only about one ten-thousandth of this is received at

Table I

	Watts transmitted	Watts received	Ratio
Power line	say 10^6	10^6	1
Submarine telegraph	5	5×10^{-7}	10^{-7}
Telephone	10^{-2}	10^{-6}	10^{-4}
Wireless telegraph...	say 10^5	10^{-8} to 10^{-12}*	10^{-13} to 10^{-17}*

* With modern thermionic apparatus.

the other end; but in both these cases the power received much exceeds that in a wireless receiver working at full commercial range. The
telephone receiver, although itself a machine of very poor efficiency,
in conjunction with the human ear is a marvellously sensitive indicator, and is used as such in wireless telegraphy. In telephony it
must reproduce intelligible speech, and this demands much greater
power than is necessary for the merely audible buzz required in
the wireless telegraph receiver. So that in wireless telegraphy the
ratio between power received and power transmitted may reach a
degree of smallness quite unapproached in telephony. Table I
shows in a rough illustrative way the orders of magnitude of the
powers transmitted and received in the several systems named.

CHAPTER II

ELECTROMAGNETIC RADIATION

1. Radiation from Electric Circuits

At a wireless station energy in the form of electromagnetic waves is radiated into space by the sending antenna, and conversely is absorbed from space by the receiving antenna. The mathematics of electromagnetic waves is beyond the scope of this book; but in the next two Sections the physical conditions will be formulated and some of the results of their mathematical analysis will be quoted. Probably in many cases, however, the reader will feel that he can only take the statements on trust, and without acquiring from them any personal conviction or insight. In this Section an attempt is therefore made to convince such a reader—to persuade him rather than to prove to him—by reference to none but familiar facts that radiation of energy must and does occur whenever the current in a circuit changes; and to show why this radiation effect is large in wireless antennae although negligibly small in our ordinary circuits. When once he apprehends that no new phenomena are involved, and that the unfamiliar differ only quantitatively from the familiar, that they are both described by the same natural laws, any irritating sense of mystery should be allayed.

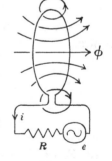

Fig. 2

Consider a circuit (Fig. 2) of resistance R carrying an alternating current $i = I \sin pt$, and producing at any instant a magnetic field such that ϕ lines thread the circuit. We are accustomed to write

$$\phi = Li = LI \sin pt$$

where L is a constant (in the absence of iron) known as the self-inductance of the circuit. It is strictly true if i changes indefinitely slowly, and it could therefore be true for rapid changes of i only if a change of current produced immediately the corresponding change of magnetic field. This would involve infinite velocity of motion of the magnetic lines—a conception sufficiently repulsive to the ordinary human being, with or without a mathematical training, for us to reject it. It follows that as i changes, the value of ϕ at any instant must diverge from the value of Li at that instant towards the value of Li during the preceding instants. As we have taken the case of a sinoidal current i, we may make some approach to this condition by supposing that ϕ is also sinoidal and lagging by some phase angle θ behind the current. Thus let us take

$$\phi = LI \sin (pt - \theta)$$

as indicated in Fig. 3.

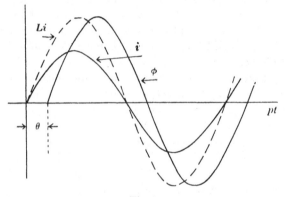

Fig. 3

By the laws of Ohm and Faraday we then have for the impressed E.M.F.

$$e = Ri + \frac{d\phi}{dt}$$

$$= RI \sin pt + pLI \cos (pt - \theta)$$

$$= \cos pt \, (pL \cos \theta) \, I + \sin pt \, (R + pL \sin \theta) \, I.$$

The first of these terms is the quadrature or reactance component of the E.M.F.; the second is the in-phase or resistance component. The total power delivered to the circuit by the alternator is obviously

$$\frac{I^2}{2} (R + pL \sin \theta).$$

Of this, $\dfrac{I^2}{2} R$ is converted into heat within the circuit, and the remainder, $\dfrac{I^2}{2}(pL \sin \theta)$. must therefore travel away from the circuit into space. $pL \sin \theta$ is thus the "radiation resistance" of the circuit; i.e. that quantity which, multiplied by the square of current, gives power radiated. We have now to see how the radiation resistance may become large.

With given p, the expression $pL \sin \theta$ for the radiation resistance is increased by increasing L or θ. L is increased by enlarging the loop or by winding on more turns of wire; and with given L, θ is increased by spreading out the field, i.e. by providing the given inductance L in a large coil of few turns rather than in a compact coil of many turns. An open antenna, consisting of a vertical straight wire one of whose ends is earthed and the other insulated and elevated, may be regarded as an extreme case giving maximum spread of field for given inductance. Again, remembering that θ is the lag of the field behind the current producing it, we see that increase of frequency will rapidly increase the radiation resistance, since both p and θ are then increased.

Similar considerations with regard to the electrostatic field instead of the magnetic field would lead to similar conclusions. We can see, therefore, that the powerful radiation from an antenna is due to the exceptionally spread-out configuration of its magnetic and electric fields, and especially to the very high frequency of the current in it.

It will be interesting at this stage to compare these rough conclusions as to conditions for marked radiation with formulae arrived at by exact mathematical analysis. M. Abraham has calculated that at a frequency of n periods per second the radiation resistance of a vertical loop consisting of T close turns of area S sq. cm is

$$7.8 \times 10^{-38} n^4 S^2 T^2 \text{ ohms.}$$

The corresponding expression for the radiation resistance of an open antenna consisting of an earthed vertical wire of length l cm is

$$7.1 \times 10^{-19} n^2 l^2 \text{ ohms.}$$

An examination of each of these expressions confirms our conclusions that the radiation resistance of a circuit increases as the magnetic or electric field is more widely spread out, and as the frequency is raised.

Several common patterns of antenna are illustrated in Fig. 24; and Plate I is a photograph of a large T antenna carried on two masts 270 feet high. An interesting contrast is provided by the small closed loop antenna of Plate II. The radiator is here a single-turn collapsible square of one-metre side, mounted on a bayonet stuck in the ground—part of a very portable military set for use in forward areas.

2. ELECTROMAGNETIC WAVES IN SPACE

Everyone is acquainted in a general way with various examples of the transmission of energy through a uniform medium by wave-motion: e.g. ripples spreading over the surface of a pond from a stone dropped in it; sound waves in the atmosphere; electric waves along lengthy power or telephone lines and along submarine cables; light waves in the aether. In these and all cases of wave-motion, the medium—e.g. the water-air surface in the case of the ripples on the pond, or the air in the case of sound—must possess two properties in the nature of mechanical elasticity and density (or their mathematical analogues), so that energy may exist at any point in the medium by virtue of its elastic deformation or of its momentum. That the medium in which electric and magnetic fields exist satisfies these conditions is recognised at once from the familiar electrostatic energy ($\frac{1}{2}Cv^2$) of the charge in a condenser and electrokinetic energy ($\frac{1}{2}Li^2$) of a current.

Now there are two fundamental laws on which the whole edifice of electromagnetic theory rests, and which in one form or another are familiar to every electrical engineer. There is Ampère's law, expressing the relation between magnetic field and the electric current producing it; and there is Faraday's law, expressing the relation between electromotive force and the changing magnetic flux producing it. Thus if ABC (Fig. 4) is any closed path

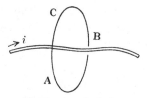

Fig. 4. Ampère's law.

linked with a circuit carrying current i, Ampère's law states that the work done on unit pole in dragging it once round the path ACB is $4\pi i$; or

$$\text{M.M.F. round ABC} = 4\pi i \quad \ldots\ldots\ldots\ldots\ldots(\text{i}).$$

And if instead of the current i we have a magnetic flux ϕ threading the closed path ABC (Fig. 5), Faraday's law states that the work done on unit electric charge in dragging it once round the path ACB is the rate of decrease of ϕ during the process (supposed uniform); or

$$\text{E.M.F. round ABC} = -\frac{d\phi}{dt} \quad \ldots\ldots\ldots\ldots\ldots\text{(ii)}.$$

The electromagnetic theory of radiation is built upon these two laws, extended by Clerk Maxwell's conception (1865) that the magnetic effect of electricity undergoing displacement in a dielectric is the same as that of an equally rapid displacement of electricity in a conductor; i.e. changing dielectric strain is the same in magnetic effect as a conductor current. Thus the conductor and its current i in Fig. 4 may be replaced by a changing dielectric strain due to an electric field whose strength F is

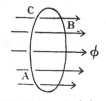

Fig. 5. Faraday's law.

changing at the rate $\dot{F}$, in which case there threads ABC, instead of conductor current i, the displacement current

$$\frac{\kappa\dot{F}}{4\pi}\,[S]_{ABC}$$

where κ = dielectric constant (specific inductive capacity) of the medium, and $[S]_{ABC}$ stands for the area of ABC resolved normally to the field*.

The M.M.F. in (i) is of course the line integral of magnetic force round ABC, and we may write it

$$[\textstyle\int H\,dr]_{ABC}$$

where H is the magnetic force along ABC at any point in it, and dr is an element of length of path ABC.

* If the field is not uniform across the area ABC, we must take elementary areas and integrate, substituting for $\dot{F}S$ the expression $\int\dot{F}\,dS$; and similarly with the analogous treatment of Faraday's law. The reader may check the above expression against the familiar case of an ordinary parallel plate condenser of capacity C, charge q, and P.D. v

$$C = \kappa \cdot \frac{S}{4\pi d}$$

$$q = vC = \frac{v}{d}\cdot\frac{\kappa S}{4\pi} = F\frac{\kappa S}{4\pi}$$

$$= \dot{q} = \dot{F}\frac{\kappa S}{4\pi}.$$

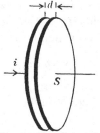

Fig. 6. Parallel plate condenser.

Equation (i) then becomes

$$[\textstyle\int H dr]_{ABC} = \kappa \dot{F}\, [S]_{ABC}$$

$$\kappa \dot{F} = \left[\frac{\int H dr}{S}\right]_{ABC} \quad \dots\dots\dots\dots\text{(iii)}.$$

Similarly equation (ii) may be rewritten

$$[\textstyle\int F dr]_{ABC} = -\frac{d}{dt}(\mu H \cdot S) = -\mu S \dot{H}$$

where μ is the magnetic permeability of the medium.

$$\therefore \quad -\mu \dot{H} = \left[\frac{\int F dr}{S}\right]_{ABC} \quad \dots\dots\dots\dots\text{(iv)}.$$

Formulae (i) and (ii) are the statements of Ampère's and Faraday's laws convenient for our everyday circuital calculations; in (iii) and (iv) we have arrived at wider statements applicable to the electromagnetic condition of space free from all conductors. If we give these two electromagnetic relations

$$\kappa \dot{F} = \frac{\int H dr}{S}$$

$$-\mu \dot{H} = \frac{\int F dr}{S}$$

to a friendly mathematician, he will, as it were by merely turning the handle of his wonderful machine, grind out for us the result

$$\frac{d^2F}{dt^2} = \frac{1}{\mu\kappa}\left(\frac{d^2F}{dx^2} + \frac{d^2F}{dy^2} + \frac{d^2F}{dz^2}\right)$$

$$\frac{d^2H}{dt^2} = \frac{1}{\mu\kappa}\left(\frac{d^2H}{dx^2} + \frac{d^2H}{dy^2} + \frac{d^2H}{dz^2}\right)$$

with the interpretation that changes of the electric and magnetic fields F and H do not make their appearance suddenly at any distant point in the medium, but are propagated as a wave-motion with the velocity $\sqrt{\dfrac{1}{\mu\kappa}}$. If we then return to our laboratory and measure μ and κ in air or vacuum, we find, using the same system of units for both quantities, that

$$\sqrt{\frac{1}{\mu\kappa}} = 3 \times 10^{10}\ \text{cm/sec.}$$

Incidentally, this velocity is the same as the best experimental determinations of the velocity of light; which is our chief ground for supposing that light is propagated through space also as an electromagnetic wave-motion, differing only in frequency from the waves of wireless telegraphy.

3. THE FIELD FAR FROM A TRANSMITTING ANTENNA

Following H. R. Hertz (1888), the father of wireless telegraphy, it is customary to endeavour to give some physical explanation of radiation from a Hertzian oscillator or an ordinary antenna by drawing a series of diagrams showing the lines of electric force at successive stages in a cycle of the alternating current in the antenna. The lines at first have ends on the two halves of the oscillator, or on the antenna and the earth, but subsequently transform themselves into self-closed loops which become independent of the antenna and travel away from it with the velocity of light. Fig. 7 shows such a diagram.

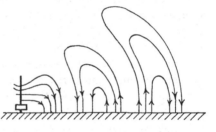

Fig. 7.   Lines of electric force.

The author believes that he is not alone in failing dismally to obtain any real enlightenment or satisfaction from these diagrams; and this presentation will not be followed here. But while the physics is, of course, contained in the mathematical equations, most people cannot grasp their meaning as a whole without resorting to some concrete picture or model based upon them. A very helpful picture of the electrical conditions in space at a distance from a radiating antenna is that developed by G. W. O. Howe in his admirable paper to the British Association at Birmingham in 1913, entitled "The nature of the electromagnetic waves employed in radio telegraphy, and the mode of their propagation*."

Howe shows that if the earth were a perfectly conducting plane surface, and there were no conducting atmosphere, the conditions near the surface of the earth at a large distance from the antenna in

* See *Electrical Review*, Sept. 26, 1913; or *Electrician*, Sept. 19, 1913.

the actual radiation case are almost exactly those in the dielectric between the two conductors of the peculiar transmission line shown in
Fig. 8. Here P is an infinite perfectly conducting plate (representing

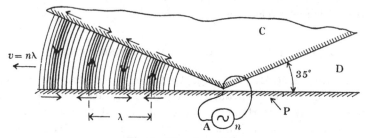

Fig. 8. Howe's model.

the surface of the earth), and C an infinite perfectly conducting cone
with axis perpendicular to P and with vertex, at which the antenna
would be located, almost touching P. D is the dielectric between
them. A is an alternator feeding current into these two conductors.
At any instant the radial conductor currents are shown in direction
by the arrows within the plate and cone, and the electric lines of
force by the lines and arrows across the dielectric between the
plate and cone. These conditions are familiar to those who have
studied telephonic currents in cables. Vacuum (or, approximately,
air) being the dielectric D, the whole system sweeps out radially
from the vertex of the cone with the velocity of 3×10^{10} cm per
sec.

At any spot near the surface of the plate (earth), the electric
field is vertical, its intensity varying harmonically with time; and
accompanying it is a horizontal magnetic field (not indicated in
Fig. 8) in phase with the electric field. At any moment at any
spot the energy per unit volume of dielectric is half electric (static)
and half magnetic (kinetic). At any point the amplitude of each
of the fields is inversely proportional to the distance from the
vertex of the cone (i.e. from the antenna). If the alternator current
in the model is equal to the current at the base of the actual simple
vertical antenna oscillating at its fundamental frequency, the same
power is propagated in the two cases provided that the angle
between cone and plane is 35°.

This model provides an excellent picture, qualitative and
quantitative, of what actually occurs in the space near the earth

around an antenna at distances exceeding several wavelengths
(so as to be sensibly beyond its non-radiative influence), and not
too far away for the effect of the curvature of the earth to be
ignored. We can use it, for example, to see how a receiving antenna
will be affected, whether an open vertical antenna influenced
mainly by the alternating electric field, or a loop antenna properly
orientated (plane containing direction of propagation) influenced
mainly by the alternating magnetic field. In Fig. 9, the upper
diagram shows over half a wavelength or so the lines of electric
force near the surface of the earth as in Fig. 8, and the lower

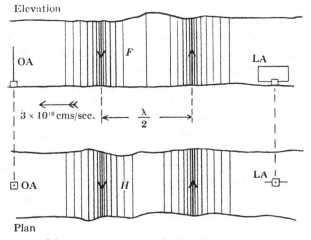

OA = open antenna (vertical single wire),
LA = loop antenna (vertical rectangle).

Fig. 9. Fields at receiving antenna.

diagram shows the lines of magnetic force at the same instant;
the regions of maximum ($+^{ve}$ and $-^{ve}$) forces being indicated by
the thick lines with arrow-heads. The whole system sweeps over
the surface of the earth in the direction of propagation (in the
plane of the paper), and produces alternating E.M.Fs. in the two
antennae shown. Both forms of antenna are affected by both
fields; but the effect of the vertical electric field on the vertical
wire OA, and the effect of the changing magnetic field threading
the loop LA, are easily grasped.

The model of Fig. 8 gives an attenuation of amplitude

proportional to distance from the transmitting antenna. Owing to curvature of the earth, at large distances the intensity actually falls off much more rapidly than as the reciprocal of the distance; and though this effect is offset (it is believed) by the presence of a conducting upper atmosphere, the "Heaviside layer," tending to retain the energy on this planet, L. W. Austin's and other experiments have shown that in daylight the intensities are proportional, not to $\frac{1}{x}$, but to an attenuation factor

$$\frac{1}{x} \times \epsilon^{-\frac{\cdot0015}{\sqrt{\lambda}}x}$$

where x is the distance from the transmitting antenna and λ is the wavelength, both in kilometres. The exponential term becomes important only when x exceeds (say) $70\sqrt{\lambda}$; e.g. with $\lambda = 1,000$ metres, when x exceeds 70 km. The attenuation given by the above semi-empirical expression is very much smaller (at large values of x) than the attenuation calculated by A. Sommerfeld and others for a conducting earth without atmosphere; and it seems probable that we are indebted to the conducting upper atmosphere for the difference.

Using the above empirical expression for attenuation, formulae have been established expressing the current in a tuned receiving antenna in terms of the current in the transmitting antenna and the various other dimensions involved*. Thus for a pair of open antennae consisting of a wire connected at the bottom to the earth and at the top to a system of conductors equivalent to a condenser plate at a height h above the earth (Fig. 10)

Fig. 10

$$I_r = I_s \times \frac{190h_sh_r}{R\lambda} \times \frac{\epsilon^{-\frac{\cdot0015}{\sqrt{\lambda}}x}}{x}$$

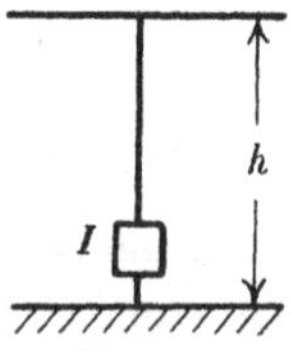

where subscripts r and s refer to the receiving and sending antennae respectively; R is resistance of receiving antenna in ohms; λ, x and h are in kilometres.

---

* Corresponding formulae are available for closed loop antennae at either or both stations. They are derived in a very important and illuminating account of the theory of radiation from antennae by J. H. Dellinger, "Principles of radio transmission and reception with antenna and coil aerials," *Scientific Paper of the Bureau of Standards*, No. 354, issued Dec. 1919.

Table II shows how, according to this formula, the received current I_r varies with the distance x for the particular conditions:

$$h_s = h_r = 100 \text{ metres} = \cdot 1 \text{ km}$$
$$\lambda = 4{,}000 \text{ metres} = 4 \text{ km}$$
$$I_s = 100 \text{ amperes}$$
$$R = 10 \text{ ohms}$$

whence
$$I_r = 4 \cdot 8 \frac{e^{-\cdot 00075\,x}}{x} \text{ amps.}$$

Table II

x kilometres	$\dfrac{4 \cdot 8}{x} \times 10^6$	I_r microamps.
10	480,000	480,000
100	48,000	45,000
300	16,000	12,800
1,000	4,800	2,300
3,000	1,600	170
4,000	1,200	60
5,000	960	22
6,000	800	9
7,000	690	3·4
8,000	600	1·5
9,000	530	·6
10,000	480	·26

With an ordinary pre-war type of receiver, the limiting range would be about 6,000 km.

It should be noticed in this table that every increment of 1,000 km in the high ranges divides the received current by approximately the same figure of 2·5. In other words, as the range is increased, in order to keep I_r to a workable value, the sending current I_s must be multiplied by 2·5 for each extension of the range by 1,000 km; the corresponding increase of power at the transmitter being, of course, $2 \cdot 5^2 = 6 \cdot 2$. Thus, to take a concrete illustration, suppose the resistance of the transmitting antenna is 2 ohms, and that for a range of 6,000 km, 100 amps. suffices. The power for 6,000 km is then $100^2 \times 2$ watts $= 20$ kW. For 7,000 km we should need $20 \times 6 \cdot 2 = 124$ kW; and for 8,000 km, 770 kW. (So large a power as this has probably never yet been delivered to an antenna.) The reason for this relationship between increase of range and the necessary increase of power is readily seen from the formula already quoted (p. 14). This may be rearranged as

$$(I_s)_x = kI_r x \epsilon^{-\frac{\cdot 0015}{\sqrt{\lambda}}x}$$

where k is a constant,

$$\therefore \ (I_s)_{x+d} = kI_r (x+d) \ \epsilon^{-\frac{\cdot 0015}{\sqrt{\lambda}}(x+d)}$$

$$= (I_s)_x \times \epsilon^{-\frac{\cdot 0015 d}{\sqrt{\lambda}}} \ \text{approx.}$$

if d is small compared with x;

$$\therefore \ \frac{(I_s)_{x+d}}{(I_s)_x} = \epsilon^{-\frac{\cdot 0015 d}{\sqrt{\lambda}}} \ .$$

Assuming that the sending current I_s and the receiving resistance R are not much affected by change in λ, the manner in which x and λ occur in the formula quoted, viz.

$$I_r = I_s \cdot \frac{190 h_s h_r}{R} \cdot \frac{\epsilon^{-\frac{\cdot 0015}{\sqrt{\lambda}}x}}{\lambda x}$$

shows that for each value of the range x there is a particular value of λ which will make I_r a maximum, viz. when

$$\frac{d}{d\lambda} \left(\frac{\epsilon^{-\frac{\cdot 0015}{\sqrt{\lambda}}x}}{\lambda} \right) = 0$$

i.e. when

$$\lambda = \frac{(\cdot 0015 x)^2}{4}$$

or λ (metres) = $\cdot 00056$ [x (kilometres)]^2.

Though there may be reasons for departing from it for other considerations, this is the optimum wavelength as far as propagation efficiency is concerned. For 6,000 kilometres the optimum wavelength is about 20,000 metres*. If this had been used instead of the 4,000 metres of Table II, the received current would have been 21 instead of 9 microamperes.

---

* The greatest wavelength in regular use is believed to be that of the arc station at Annapolis, U.S.A.; about 17,000 metres.

PLATE III. FIXED SPARK GAP (p. 40).
(Marconi's Wireless Telegraph Co., Ltd.)

PLATE IV. GLASS-PLATE TRANSMITTING CONDENSER (p. 40).
(Marconi's Wireless Telegraph Co., Ltd.)

It must be added that the empirical attenuation factor, $\frac{1}{x} \times \epsilon^{-\frac{\cdot0015}{\sqrt{\lambda}}x}$ (Austin-Cohen), has not been securely established as of general validity. It has been checked with ranges up to about 4,000 kilometres and with various wavelengths up to about 4,000 metres; but recent observations by G. Vallauri on the signals from Annapolis ($\lambda = 17,300$ m.) received at Leghorn (distant about 7,000 km) are much better fitted by an attenuation factor suggested in 1915 by L. F. Fuller, $\frac{1}{x} \times \epsilon^{-\frac{\cdot0045}{\lambda^{1\cdot4}}x}$ *. This leads to an optimum wavelength

$$\lambda \text{ (metres)} = 27\cdot5 \, [x \text{ (kilometres)}]^{\cdot71}$$

which is about 13,000 metres when $x = 6,000$ kilometres. With the Fuller factor in place of the Austin-Cohen factor, and at the optimum wavelength, the received current at 6,000 kilometres in Table II would have been 125 microamperes, instead of the tabulated 9 microamperes or the Austin-Cohen optimum 21 microamperes. It is clear that further experiments are needed to establish an attenuation factor valid for all ranges and wavelengths.

* See G. W. O. Howe, "Measurement of the field strength at Leghorn of the Annapolis signals." *Radio Review*, Oct. 1920.

CHAPTER III

OSCILLATORY CIRCUITS

1. REACTANCES AND RESISTANCES

As already stated, when we leave the subject of radiation and come to consider the sensibly non-radiating circuits in which high-frequency currents are generated, transformed and controlled, we are amongst the phenomena of more ordinary alternating current engineering, with only such quantitative differences as derive from the enormously high frequency of the currents. These circuits are used at the transmitter in conveying power to the antenna for producing electromagnetic waves in space, and at the receiver in withdrawing power from the antenna and passing it to some form of detector.

From a variety of circumstances, in wireless circuits the reactances are usually very large compared with the resistances. Sometimes this condition cannot be avoided, as in antennae; sometimes it is convenient for the production with small power of the large E.M.F. required by accessory apparatus such as crystal detectors and triodes; sometimes, too, it is cultivated with the object of filtering out one frequency from another, an operation requiring oscillatory circuits of low damping. Since it is only the resistance component of the E.M.F. which is involved in the work done with which we are ultimately concerned, it is therefore necessary to counter an inductive E.M.F. by an equal and opposite capacitative E.M.F. In other words, wireless circuits are generally—not invariably—tuned oscillatory circuits, comprising inductance and capacity so adjusted that their reactances are approximately numerically equal at the particular frequency employed, each of these reactances alone very greatly exceeding the resistance.

2. Undamped impressed E.M.F.

The equation of E.M.Fs. in the circuit of Fig. 11 is

$$L\frac{di}{dt} + Ri + \frac{1}{C}\int idt = e = E \sin 2\pi nt$$

$$\therefore \ L\frac{d^2i}{dt^2} + R\frac{di}{dt} + \frac{1}{C}i = 2\pi n \, E \cos 2\pi nt.$$

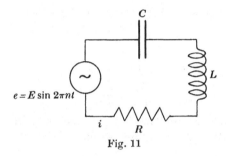

$e = E \sin 2\pi nt$

Fig. 11

The solution of this equation is

$$i = I \sin (2\pi nt - \theta) + I' \epsilon^{-bt} \sin (2\pi n't - \phi)*$$

where

$$I = \frac{E}{\sqrt{\left(2\pi nL - \dfrac{1}{2\pi nC}\right)^2 + R^2}}$$

$$b = \frac{R}{2L}$$

$$n' = \frac{1}{2\pi}\sqrt{\frac{1}{LC} - \frac{R^2}{4L^2}}.$$

* An important particular case is when

$$2\pi n = \frac{1}{\sqrt{LC}}$$

$$\frac{R^2}{4L^2} << \frac{1}{LC}.$$

If the condenser holds no initial charge and the alternator is switched into circuit at $t = 0$, the solution then becomes approximately

$$i = \frac{E}{R} \sin pt \, (1 - \epsilon^{-\frac{R}{2L}t}).$$

From this we see that:

(i) There is a free oscillation of frequency

$$n' = \frac{1}{2\pi} \sqrt{\frac{1}{LC} - \frac{R^2}{4L^2}}$$

which usually in wireless circuits is sensibly equal to

$$\frac{1}{2\pi\sqrt{LC}}.$$

(ii) This free oscillation has an attenuation constant

$$b = \frac{R}{2L}$$

and therefore a logarithmic decrement*

$$\delta = \frac{b}{n'} = \frac{R}{2n'L}.$$

(iii) After the free oscillation has died away there remains only the forced oscillation of amplitude

$$I = \frac{E}{\sqrt{\left(2\pi nL - \frac{1}{2\pi nC}\right)^2 + R^2}}.$$

(iv) This is a maximum when the circuit is in resonance with the frequency n of the impressed E.M.F. e, i.e. when

$$2\pi nL = \frac{1}{2\pi nC}$$

i.e. when

$$\text{forced period } n = \frac{1}{2\pi\sqrt{LC}} = \text{free period } n' \text{ approx.}$$

$$= \frac{p}{2\pi} \text{ (say).}$$

(v) The current amplitude is then $\frac{E}{R}$: i.e. the circuit behaves as though its only impedance were its resistance R.

---

* Commonly abbreviated to "decrement." The decrement δ is generally a more convenient measure of damping than is the attenuation constant b because it can be used without reference to frequency. Thus, the amplitude of an oscillatory system, when disturbed and left to oscillate freely, is divided by $\epsilon \approx 2\cdot7$ during the course of every $\frac{1}{\delta}$ cycles, and approximately dies out altogether in (say) $\frac{4}{\delta}$ cycles.

Since pL and $\dfrac{1}{pC}$ are usually very large compared with R, slight inequalities in pL and $\dfrac{1}{pC}$, i.e. slight departures from resonance, make great changes in the values of current given by the expression in (iii). As an example, in Fig. 12 is plotted a curve*, commonly called a "resonance curve," between I and n for a lightly damped circuit such as might occur in practice, where

$$R = 10\,\Omega \qquad L = 5{,}000\mu\mathrm{H} \qquad C = 508\mu\mu\mathrm{F}$$
$$\lambda = 3{,}000\,\mathrm{m} \qquad n' = 10^5\ \mathrm{p.p.s.} \qquad \delta = \cdot01$$

and taking $E = 10$ volts.

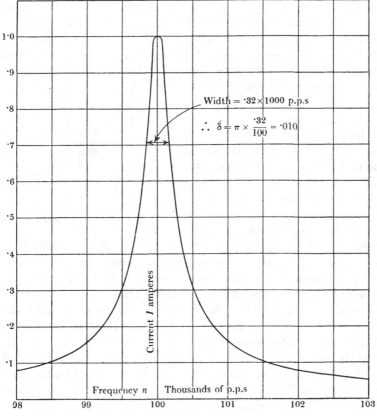

Fig. 12. Resonance curve.

* See also curves in Fig. 92.

The "peakiness" of this curve measures the "sharpness of tuning"; and it may be shown (V. Bjerknes) that the decrement, if small, is π times the width of the peak at $\dfrac{1}{\sqrt{2}}$ times its height, divided by the abscissa of the peak. This remains true if the ordinates represent R.M.S. instead of maximum values of current. Decrements are conveniently determined in the laboratory in this way.

3. THE WAVEMETER

It is obvious that if the circuit of Fig. 11 is provided with some form of current measurer such as a hot-wire ammeter inserted in the circuit, and if the frequency n of the impressed E.M.F. is varied or if the natural frequency n' of the circuit is varied by variation of L or C, the ammeter will indicate when $n = n'$ by showing then its maximum deflection. The values of L and C being variable and known, the LC circuit then constitutes a frequency-meter, which is the more sensitive as its decrement is reduced.

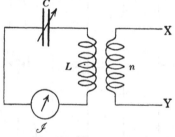

Fig. 13. Wavemeter.

In Fig. 11 the E.M.F. is shown as impressed upon the circuit by actually connecting the generator in series with it. A more common condition is that in which the E.M.F. is induced in the inductance L by placing it in the magnetic field of a high-frequency current, as in Fig. 13. Provided that the reaction of the frequency-meter circuit on the circuit XY, the frequency n of whose current is to be measured, is small enough not appreciably to affect that frequency, we can determine n by varying the calibrated condenser C until the observed current $\mathcal{J}$ is a maximum. A calibrated circuit of this kind is commonly called a "wavemeter," the graduations on the scale being generally the wavelengths (in metres) which correspond with the particular natural frequency of the circuit at that adjustment. It is in fact customary to speak of the wavelengths of high-frequency currents even when no waves are contemplated: what is signified is, of course, the wavelength of the radiation in space which would correspond with the particular frequency under consideration. The convenience of this custom arises from the fact that we are seldom able directly to count the number of periods per second, whereas we do constantly measure lengths of waves more or less directly, and perform gross operations such as cutting a length of wire for an antenna equal to some large fraction (e.g. $\frac{1}{4}$) of the wavelength. Wavelength, in this sense, and frequency are always interchangeable according to the relation

$$\lambda \times n = 3 \times 10^{10} \text{ cm per sec.}$$

It is convenient to remember that 300 metres, one of the wavelengths allocated for ships by International Convention, corresponds with a frequency of one million periods per second.

A wavemeter provided with an instrument to *measure* the R.M.S. current, as in Fig. 13, can be used to obtain resonance curves as in Fig. 12. But any indicator capable of showing merely when the current is a maximum suffices for measuring the wavelength. A wavemeter of this simpler form is as indispensable from hour to hour to the wireless man as his detector (galvanometer) is to the lineman.

4. DAMPED IMPRESSED E.M.F., AND COUPLED CIRCUITS

If a charged condenser is discharged through a circuit possessing inductance (and not too large resistance), a damped oscillation takes place. In Fig. 14, while the switch S is open the condenser remains charged to a P.D. V; and when S is closed there is an oscillatory discharge in $CLRS$ for which the equation of E.M.Fs. is*

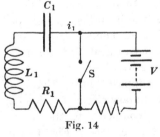

Fig. 14

$$L_1 \frac{di_1}{dt} + R_1 i_1 + \frac{1}{C_1} \int i_1 dt - V = 0$$

$$\therefore \; L_1 \frac{d^2 i_1}{dt^2} + R_1 \frac{di_1}{dt} + \frac{1}{C_1} i_1 = 0.$$

The accurate solution of this equation is

$$i_1 = V \frac{C_1 (p_1{}^2 + b_1{}^2)}{p_1} \epsilon^{-b_1 t} \sin p_1 t$$

* Cf. p. 19. Here there is no forced oscillation.

where
$$b_1 = \frac{R_1}{2L_1}$$

$$p_1 = \sqrt{\frac{1}{L_1 C_1} - \frac{R_1{}^2}{4L_1{}^2}}$$

and t is measured from the moment of closing of the switch. With slight damping, i.e. $b_1 < < p_1$, this simplifies to

$$i_1 = I_1 \epsilon^{-b_1 t} \sin p_1 t$$

with $I_1 = \frac{V}{p_1 L_1}$ and $p_1 = \sqrt{\frac{1}{L_1 C_1}}$ approximately.

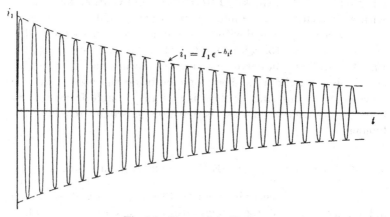

$i_1 = I_1 \epsilon^{-b_1 t}$

Fig. 15. Damped oscillation.

Fig. 15 shows the oscillation dying away asymptotically to zero as the initial energy of the charged condenser is dissipated in heating the resistance R_1.

Suppose now that a secondary circuit C_2, L_2, R_2 is "coupled" to the oscillatory circuit of Fig. 14 by the existence of a mutual inductance M between L_1

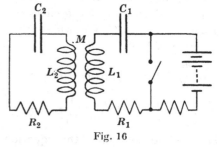

Fig. 16

and L_2, as in Fig. 16. If the coupling is sufficiently weak to allow us to ignore the reaction E.M.F. $M \frac{di_2}{dt}$ in the primary circuit, the oscillation produced in the secondary is clearly

$$L_2 \frac{di_2}{dt} + R_2 i_2 + \frac{1}{C_2} \int i_2 dt = -M \frac{di_1}{dt} = -MI_1 \frac{d}{dt}(\epsilon^{-b_1 t} \sin p_1 t)$$

$$\therefore \ L_2 \frac{d^2 i_2}{dt^2} + R_2 \frac{di_2}{dt} + \frac{1}{C_2} i_2 = -MI_1 \frac{d^2}{dt^2}(\epsilon^{-b_1 t} \sin p_1 t)$$

$$= p^2 MI_1 \epsilon^{-b_1 t} \sin p_1 t$$

approximately, when primary damping is not great.

If neither of the two circuits is extravagantly damped, the secondary current is greatest when the two circuits are "tuned" together*. Then the solution of the equation is approximately

$$i_2 = \frac{-pMI_1}{2L_2(b_2 - b_1)}(\epsilon^{-b_1 t} - \epsilon^{-b_2 t}) \sin pt.$$

This expression may be rewritten in a form showing the process of building up and decay more clearly in terms of the number of periods N (instead of time t) from the beginning, by making the substitutions

$$p = 2\pi n$$
$$pt = 2\pi N$$
$$b = n \cdot \delta$$
$$bt = N \cdot \delta.$$

Then
$$i_2 = \frac{-\pi MI_1}{L_2(\delta_2 - \delta_1)}(\epsilon^{-\delta_1 N} - \epsilon^{-\delta_2 N}) \sin 2\pi N.$$

Fig. 17 shows the course of an oscillation of this character, together with the primary oscillation producing it, for the decrements

$$\delta_1 = \cdot 06$$
$$\delta_2 = \cdot 20.$$

In Fig. 17, the secondary circuit has a rather high decrement ($\cdot 2$). If a wavemeter is substituted for it, and a resonance curve is taken (as in Fig. 12 except that ordinates are R.M.S. instead of maximum values of current), the peakiness of this curve will depend on the sum of the primary and secondary decrements ($\delta_1 + \delta_2$). It may be proved that the width shown in Fig. 12 as giving δ will now give ($\delta_1 + \delta_2$), from which δ_1 is found if δ_2, the decrement

---

* Strictly, two oscillatory circuits are said to be "tuned" or "isochronous" when $\left(\frac{1}{LC} - \frac{R^2}{4L^2}\right)$ is the same in each, and to be "resonant" when $\frac{1}{LC}$ is the same in each. In practice, the damping is usually so small that no distinction is necessary, and "tuned" is commonly used when "resonant" would be strictly correct.

of the wavemeter, is known or negligibly small in comparison. In order to obtain such a curve in practice it is necessary to produce a continued regular succession of primary discharges by repeatedly and regularly opening and closing the switch in Fig. 16, so that a steady reading of the ammeter in the wavemeter circuit is obtained for each wavelength adjustment.

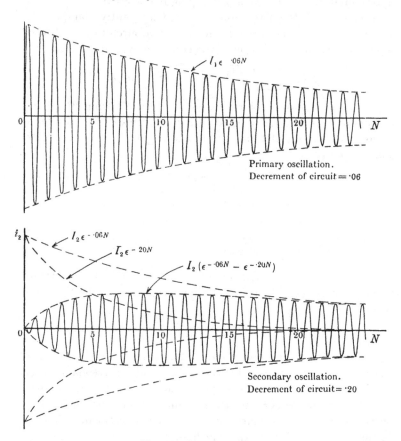

Primary oscillation.
Decrement of circuit = ·06

Secondary oscillation.
Decrement of circuit = ·20

Fig. 17. Loose coupling.

If the two circuits we have been considering (Fig. 16) are more closely coupled together by increasing their mutual inductance M, the reaction E.M.F. $M \dfrac{di_2}{dt}$ becomes great enough to affect appreciably the primary current, and some of the energy acquired by the secondary is subsequently returned to the primary*. If the two circuits are tuned, this interchange of activity is complete, and with small dampings may occur repeatedly before all the energy is dissipated. Fig. 18 shows the alternate waxing and waning of the primary and secondary amplitudes in a pair of tuned coupled circuits oscillating in this manner.

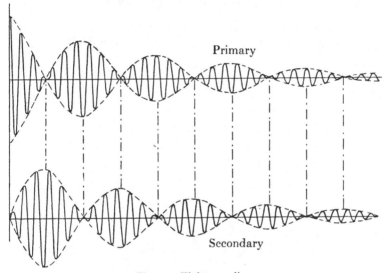

Fig. 18. Tight coupling.

The precise shapes of the curves of current or P.D. (P.D. in Fig. 18) need not concern us here. The dotted envelopes, showing the changing amplitudes of successive cycles, are the important feature, and indicate how the energy of the oscillation repeatedly

* This is not an accurate statement, as the interchange must depend on the starting conditions. For example, it is obvious that if the two circuits were similar in all respects and at some instant contained oscillations equal in amplitude and exactly in or exactly out of phase, no transference would take place; for any argument for transference of energy from one to the other would be equally cogent for transference from the other to the one. Transference requires a phase angle of magnitude greater than 0° or less than 180° (according to the direction of transference) between the current in one circuit and the E.M.F. induced in it by the other current. Taking account of damping the mathematical analysis is very complicated, even in special cases, and probably most engineers will find a few experiments with a simple mechanical model more illuminating. (See Appendix, p. 34.) For a summary of the mathematical treatment the student should consult J. A. Fleming's treatise, *The Principles of Electric Wave Telegraphy* (3rd ed. pp. 298–336); but he must beware of hidden approximations and assumptions.

changes its seat from one circuit to the other. At any moment in either circuit there exists an alternating current of relatively slowly fluctuating amplitude, recalling the familiar beat effect of two tones of nearly equal frequencies in acoustics. Provided that the dampings are small it may be shown if n_1, n_2 are the natural frequencies of the primary and secondary circuits when separated, the coupled system has two natural frequencies n_a, n_b given by

$$n_a{}^2 \text{ or } n_b{}^2 = \frac{2n_1{}^2n_2{}^2}{n_1{}^2 + n_2{}^2 \pm \{(n_1{}^2 - n_2{}^2)^2 + 4k^2n_1{}^2n_2{}^2\}^{\frac{1}{2}}},$$

where $k \equiv$ "coefficient of coupling" $\equiv \dfrac{M}{\sqrt{L_1 L_2}}$. An examination of this expression shows that if n_1 is the smaller of the two natural frequencies n_1 and n_2, then $n_a < n_1$ and $n_b > n_2$.

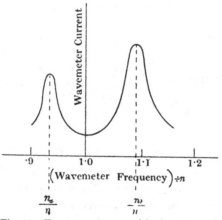

Fig. 19. Resonance curve with close coupling.

If the two circuits are tuned, i.e. if $n_1 = n_2 = n$, the expression simplifies to

$$n_a = \frac{n}{\sqrt{1 + k}}$$

$$n_b = \frac{n}{\sqrt{1 - k}}.$$

The beat effect exhibited in Fig. 18 may be regarded as due to the interference between two coexisting currents of these frequencies in each of the circuits. The frequency of the beats is the difference of the two constituent frequencies, i.e. $(n_a - n_b)$.

The oscillation in either circuit may be examined by taking a resonance curve with a wavemeter loosely coupled to it, as already explained. The existence of the two frequencies is then revealed by the appearance of two peaks in the resonance curve. Fig. 19 shows such a curve and corresponds to the oscillations of Fig. 18.

As the mutual inductance M, and therefore the coefficient of coupling $k = \dfrac{M}{\sqrt{L_1 L_2}}$, between the two circuits is gradually reduced, the two peaks of the resonance curve draw gradually together, until when the coupling is very "light" or "loose" they coalesce. We have then once more the case of indefinitely light coupling portrayed in Fig. 17.

5. METHODS OF COUPLING CIRCUITS

The coupling together of two oscillatory circuits consists, as we have seen, in disposing them in such a way that an alternating current in one circuit impresses an alternating E.M.F. in the other. In the foregoing we have examined the particular form of coupling which is provided by the existence of a mutual inductance between the two circuits (Fig. 16). In this case, if the current in one circuit is

$$i_1 = I_1 \sin pt$$

the E.M.F. introduced into the other circuit is

$$e_2 = M \frac{di_1}{dt} = pMI_1 \cos pt.$$

This form of coupling is of the commonest occurrence, and possesses the two advantages that there is no conductive connection between the two circuits*, and that the closeness of the coupling is easily variable without any incidental alteration of the natural frequencies of the individual circuits. Thus in Fig. 16, the inductances L_1, L_2 may be those of two mechanically separate coils of wire wound on separate frames (e.g. ebonite or cardboard cylinders, one sliding into the other), and their mutual inductance M may be varied by merely varying their relative position. Fig. 20 shows a convenient form of construction, in which one coil can be rotated within the other, so that M varies from zero when their planes are perpendicular to a maximum when their planes are coincident.

* This is often a great practical convenience, particularly where alternating and direct currents and E.M.Fs. are superposed in the same circuit, as in triode receiving circuits (Chapter VIII).

Precisely the same kind of interaction between the alternating currents in the two circuits is produced when they possess a common

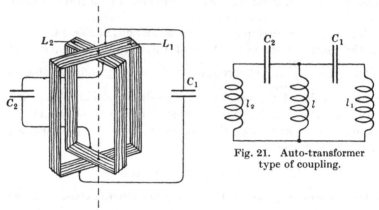

Fig. 21.   Auto-transformer
type of coupling.

Fig. 20.   Variable mutual inductance.

self-inductance instead of a mutual inductance. Thus in Fig. 21, a primary current

$$i_1 = I_1 \sin p_1 t$$

impresses an E.M.F.

$$e_2 = l\frac{di_1}{dt} = plI_1 \cos p_1 t$$

in the secondary circuit; and if we write $(l_1 + l)$ for L_1, $(l_2 + l)$ for L_2, and l for M the previous equations stand unchanged, and the coefficient of coupling is obviously

$$k = \frac{l}{\sqrt{(l_1 + l)\,(l_2 + l)}}.$$

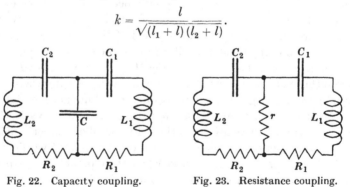

Fig. 22. Capacity coupling.   Fig. 23.   Resistance coupling.

Figs. 22 and 23 show. examples of capacity coupling and resistance coupling, respectively.   Resistance coupling is not of much practical importance, but capacity coupling often occurs. In Fig. 22, a primary current

$$i_1 = I_1 \sin p_1 t$$

impresses an E.M.F. $\dfrac{1}{C}\displaystyle\int i_1 dt$

in the secondary circuit; so that

$$L_2 \frac{di_2}{dt} + R_2 i_2 + \frac{1}{C_2}\int i_2 dt = \frac{1}{C}\int i_1 dt$$

$$\therefore\ L_2 \frac{d^2 i_2}{dt^2} + R_2 \frac{di_2}{dt} + \frac{1}{C_2} i_2 = \frac{1}{C} I_1 \sin p_1 t.$$

Comparing this with the corresponding equation on p. 25 for coupling by mutual inductance M, we see that they are the same if we write $\dfrac{1}{C}$ in place of $p_1^2 M$. The coefficient of coupling is therefore

$$k = \frac{\frac{1}{p^2 C}}{\sqrt{L_1 L_2}} = \sqrt{\frac{C_1 C_2}{(C_1 + C)(C_2 + C)}}.$$

6. THE ANTENNA AS OSCILLATORY CIRCUIT

An antenna consists commonly of a lengthy wire or group of wires, supported more or less high in the air at one end and connected to earth through an inductance at the other. Several patterns of antenna are shown in Fig. 24. It often happens that one of a pair of coupled circuits (Fig. 16) consists of an antenna circuit, as in Fig. 25*. The secondary (antenna) circuit of Fig. 25 differs from the secondary circuit of Fig. 16 in the following respects:

(i) The inductance is not all concentrated in the coil L_2, but is in part distributed along the antenna itself.

(ii) The capacity is not concentrated in a compact condenser C_2, but is distributed along the antenna.

(iii) The current is not uniform throughout the circuit, but varies from a maximum at the earth connection to zero at the tips of the antenna.

(iv) There is large loss of power from the circuit by radiation, and also by currents in the ground near the antenna, the soil being an imperfect dielectric.

* The secondary decrement δ_2 in Fig. 17 was, in fact, chosen as that of a rather highly damped antenna.

Nevertheless, for any particular small range of frequencies, the actual antenna circuit in its local effects is very nearly equivalent to a closed, i.e. substantially non-radiating, circuit with concen-

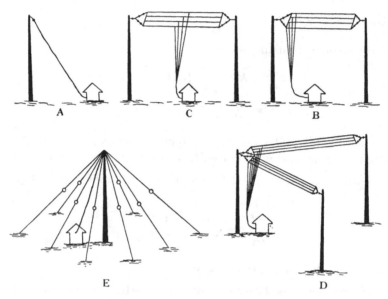

Fig. 24. Common patterns of antenna.

trated capacity C_2, inductance L_2 and resistance R_2 (as in Fig. 16), provided that C_2, L_2 and R_2 are chosen to satisfy the following conditions:

(a) $\dfrac{1}{2\pi\sqrt{C_2L_2}}$ equals the natural frequency of antenna circuit. This determines the product C_2L_2.

(b) The "equivalent concentrated circuit" and the actual antenna circuit are equally changed in frequency by the insertion of a small extra inductance. With (a) this determines the values of C_2 and L_2 individually.

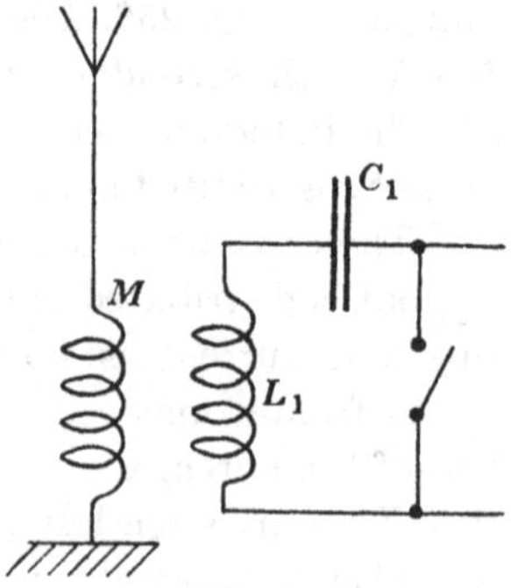

(c) $\mathscr{I}_2{}^2R_2$ equals the power dissi-

Fig. 25

pated in the actual antenna, when the R.M.S. current at some specified point in it—invariably taken as the bottom—is also $\mathscr{I}_2$,

PLATE V. DISC DISCHARGER, 60 KW (p. 41).
(Marconi's Wireless Telegraph Co., Ltd.)

PLATE VI. MULTIPLE QUENCHED SPARK GAP (p. 43).
(Gesellschaft für drahtlose Telegraphie m. b. H.)

the R.M.S. current in the equivalent concentrated circuit. This determines R_2.

Whenever we speak without qualification of the capacity, inductance, or resistance of an antenna circuit, we are tacitly referring to the equivalent concentrated circuit; just as when we speak of the length of a pendulum we really mean the length of the ideal simple equivalent pendulum. It should be remembered, moreover, that the equivalence is a limited one, and strictly refers only to an indefinitely narrow range of frequencies about the particular frequency for which the above conditions (a), (b) and (c) are satisfied.

In an antenna of the form shown at (A) in Fig. 24, consisting of a single wire, the distribution of inductance and capacity is very thorough. The equivalent concentrated capacity of such an antenna is a minimum for oscillations of the fundamental frequency (i.e. with no loading inductance inserted in the antenna). Fig. 112 B shows how the amplitude of potential is then distributed along the antenna. As the wavelength is increased by the insertion of loading inductance, the potential distribution changes as shown at Fig. 112 C, and the capacity increases, at first rapidly. Finally, for wavelengths very much greater than the fundamental, the potential is sensibly uniform along the antenna, and the capacity becomes indistinguishable from the electrostatic capacity of the antenna when severed from earth at the bottom. At the other extreme, in an umbrella antenna (Fig. 24 E) the inductance is fairly well concentrated in the uplead, and the capacity in the radial wires at the top; the equivalent concentrated capacity is therefore fairly independent of the wavelength even when the antenna is but little loaded.

APPENDIX

MECHANICAL MODEL OF COUPLED OSCILLATORY CIRCUITS

The mathematical analysis of the behaviour of coupled circuits is either so complicated or so incomplete, that a simple mechanical model with which one can actually experiment is of great value in throwing light on the phenomena, and even in obtaining solutions to numerical problems.

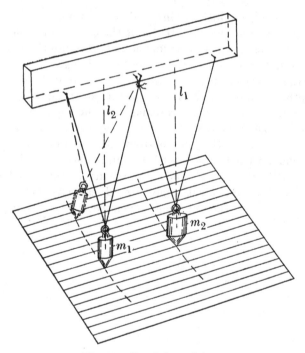

Fig. 26. Coupled pendulums.

Fig. 26 shows a pendulum model, very easily constructed, with which a great variety of experiments can be made, displacement of pendulum bob in the model representing value of current in the oscillatory circuit. The pendulums may be about a metre long, and when they represent circuits accurately tuned together, each can conveniently be adjusted to have a period of exactly two seconds. Since the ratio $\frac{L}{C}$ is proportional in the model to m^2, a change of $\frac{L}{C}$ without change of LC is readily represented. More or less damping can be added by means of paper wind-vanes fixed to the strings. The coupling between the two oscillators is increased by lowering the point where the two strings are tied together, being zero when they are not so tied provided that the supporting beam is sensibly rigid. The model departs from being an exact mathematical analogue of the electrical circuits examined in Chapter III, in that the type of coupling is more nearly that given by a common resistance than a common inductance; but as long as we are concerning ourselves with such damping and coupling as are usually found in wireless currents, this discrepancy is of small importance. If the bob of one pendulum is made very massive compared with the other, and is lightly damped, a very close approximation to a steady (undamped) primary oscillation is obtained.

CHAPTER IV

PRODUCTION OF HIGH-FREQUENCY ALTERNATING CURRENTS

1. SPARK METHODS

UNDER ordinary conditions, air and other gases are almost perfect insulators; but when subjected to sufficiently intense electric stress a gas becomes ionised and allows a current to flow. High temperature of the gas, and the projection into it of electrons emitted from a hot solid body or otherwise, tend to produce or maintain the ionisation. If therefore the P.D. between two electrodes spaced a short distance apart in air is gradually raised from a low value, at first no current flows between them; but when the P.D. reaches a value so great that the electric field is strong enough to ionise the gas, a current begins to flow, and the effect of this current is itself to increase the conductance of the air gap. If we call the ratio between the P.D. and the current across the gap the resistance of the gap, we see that the resistance is a function of the current, decreasing as the current increases*. As long as the current continues, the state of ionisation is maintained despite the cessation of the intense electric field required to initiate the ionisation. Thus in order to start a discharge across a $\frac{1}{4}''$ gap between copper electrodes, a P.D. of nearly 20,000 volts might be needed; but once started, a current of 10 amperes would be maintained by a P.D. of only about 50 volts. If the current ceases, even for a very brief period, the ionisation disappears, and the insulating properties of the gap are recovered.

A spark gap may therefore be used as a form of high-tension switch which, without moving parts, closes automatically when the P.D. across it rises to the disruptive value, and opens again when the current through the spark becomes too small to maintain the ionisation on which the conductivity of the gap depends. Thus a spark gap may replace the mechanical switch S in Fig. 14.

* Hence the familiar "negative resistance" of the ordinary electric arc, which can only be fed from a source of constant P.D. if the circuit is stabilised by the insertion of a ballast resistance.

Provided that the P.D. V of the battery or other source charging the condenser C_1 is high enough to jump the spark gap, as C_1 charges up the P.D. across the gap gradually rises towards the value V. When it reaches the disruptive value* the gap breaks down, and its resistance falls from a sensibly infinite value to some low value of the order of an ohm. The condenser C_1 thereupon discharges through L_1R_1 and the gap with a damped oscillation very much as in Fig. 15†. When the oscillation has died down to a very small amplitude, there is nothing to maintain ionisation, and the insulating property of the gap is restored. When the condenser is again sufficiently charged up, a new spark occurs, and the whole process repeats itself indefinitely.

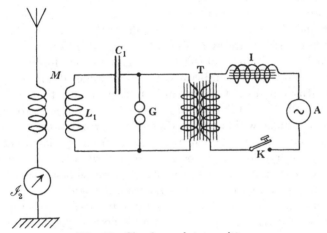

Fig. 27. Simple spark transmitter.

If the battery and series resistance are replaced by the more convenient spark coil, or alternator with step-up transformer, we have the commonest of all forms of generator of high-frequency current. Indeed, until a few years ago practically all wireless

* The sparking P.D. depends somewhat on the shape of the electrodes, and is roughly proportional to the length of the gap. For air under ordinary atmospheric conditions, the P.D. necessary to jump a ¼″ gap varies from about 8,000 volts between sharp needle points to about 20,000 volts between large smooth spheres.

† Any difference is due to the fact that the resistance of the gap is not constant, but depends upon the current flowing, and on other factors determining the state of ionisation. The amplitude therefore does not fall off strictly according to the exponential law of Fig. 15; but this forms a sufficiently close approximation for our present purposes.

transmitters were of this type*, and even to-day probably well over 95 per cent. are spark transmitters. The circuits of a typical simple spark transmitter are shown in Fig. 27. A is a low-frequency alternator (of say 50–500 periods per second) connected at will through the key K to the step-up transformer T which provides current at high tension to charge the condenser C_1. When a spark occurs across the gap G, the high-frequency oscillation in the spark circuit L_1C_1G induces a powerful oscillation in the tuned antenna (secondary) circuit by virtue of the mutual inductance M between the two circuits. Energy is radiated from the antenna in all directions over the surface of the earth at a rate which is proportional to the square of the current $\mathscr{I}_2$ indicated by a hot-wire ammeter inserted at the base of the antenna circuit.

According as the coefficient of coupling† between the spark circuit and the antenna circuit is small or large, the oscillations in the two circuits are as in Fig. 17 or Fig. 18 respectively. The condition of Fig. 18 is objectionable, and in fact forbidden by International Convention, because the radiation contains waves of two lengths, only one of which can be fully utilised at the receiving station, while the other may be an unnecessary nuisance to other receiving stations within range. On the other hand, if the coupling is very loose, the transference of energy from spark circuit to antenna is very slow, and poor efficiency results; for as long as oscillation continues in the spark circuit power is being wasted in the spark gap, whose resistance is far from negligible. Thus to get rid of the energy quickly tight coupling is desired; but to keep the singleness of wavelength loose coupling is necessary. The compromise adopted may be a coefficient of coupling $\dfrac{M}{\sqrt{L_1L_2}}$ in the region of 6 per cent.; in which case, if two wavelengths were detectable at all, they would be about 97 per cent. and 103 per cent. of the common wavelength of the two circuits when separated. Actually, however, with the coefficient of coupling used, the antenna circuit generally has a decrement high enough to prevent two peaks from showing in the resonance curve.

As soon as the oscillation has died away in the spark circuit—

---

* Witness the German "Funkentelegraphie," which was almost synonymous with "wireless telegraphy"; and "Telefunken" Company as the colloquial name for the Gesselschaft für drahtlose Telegraphie.

† See p. 28.

e.g. in $\frac{1}{10,000}$ second or so—the gap de-ionises and is again an insulator; the condenser C_1 (Fig. 27) receives a fresh charge from the alternator; and the process repeats itself as long as the key is held depressed. The Morse dots and dashes thus consist of short or long series of sparks, each spark setting up in the antenna a brief and violent oscillation followed by a relatively lengthy period of quiescence before the next spark occurs. The rate at which the sparks follow each other depends on the time taken to recharge the condenser to the sparking voltage, and this depends on the alternator frequency and on the length of the gap. The older sets make use of alternators of ordinary commercial frequency such as 50 p.p.s.; the spark gap is then set so that it breaks down at a voltage much less than the maximum which would be produced during a non-sparking cycle, and several sparks occur at irregular intervals during the half-cycle.

A trouble which is liable to occur is the formation of an arc across the spark gap; that is, after the gap has been rendered conducting by the disruptive spark and high-frequency oscillation, an arc may be established, fed directly from the transformer and alternator, thus wasting power and preventing the recovery of the insulating property of the gap. The iron-cored inductance I shown in Fig. 27 inserted between alternator and transformer tends to inhibit such arcing by retarding the re-charge of the condenser during and immediately after the oscillatory discharge through the spark gap. An additional measure against arcing which is sometimes adopted is to use a rotary spark gap, in which the electrodes are made to approach and separate at high speed several times per half-period of the alternator. A rotary gap tears out any incipient arc, and the incidental ventilation and cooling tend to prevent any arc from forming. Such a gap, moreover, tends to equalise the intervals between successive sparks.

Some notion of practical dimensions in a simple spark transmitter of this kind may be got from the following figures referring to a transmitter such as might be found on a small ship or shore station for commercial work.

Wavelength = 600 metres.   Frequency = 5×10^5 p.p.s.
Alternator output = 2 kilowatts at 50 p.p.s.
Mean spark rate = 500 per sec.
Sparking P.D. = 10,000 volts. C_1 = ·05 microfarad.
Antenna current = 6 amps. (R.M.S.).

Plate III shows a form of fixed spark gap which might be used in such a station; and Plate IV a common type of high-tension condenser C_1. This condenser is constructed of alternate glass and zinc plates in a galvanised iron tank filled with oil. In transmitters of low power, mica condensers sealed in wax are often used.

In a more modern form of spark transmitter, the spark takes place at a rotary gap mounted on the shaft of the alternator, as illustrated diagrammatically in Fig. 28. By making the number of moving electrodes equal to the number of poles, one spark per half-

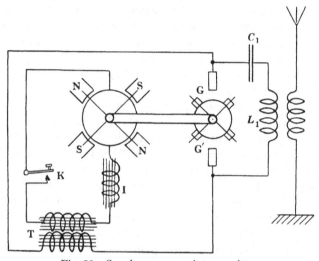

Fig. 28. Synchronous spark transmitter.

cycle is obtained; and by adjusting the position of the stationary electrodes the spark is made to occur approximately at the moment when the transformer current is zero and the condenser P.D. a maximum. The alternator frequency is made half the desired spark frequency, and is commonly 200–500 p.p.s. The iron-core inductance I is adjusted to give fairly precise low-frequency resonance, i.e. to make the natural frequency of the compound circuit $AITC_1$ (Fig. 27) approximately equal to the alternator frequency. By these means, precisely regular sparking is obtained, quite free from arcing, and with a high power-factor at the alternator terminals. Fig. 29* depicts a low-frequency half-cycle under these conditions; the

* From "The low-frequency circuit in spark telegraphy," by L. B. Turner, *Electrician*, Aug. 2, 1912.

spark is there supposed to occur exactly at $pt = 180°$, whereupon the whole process repeats itself in the reverse direction.

This system is known as the synchronous spark system; it has been very widely used by the Marconi Co., for powers between $\frac{1}{2}$ kW and upwards of 100 kW. Plate V shows a large (60 kW) disc discharger which would be rigidly coupled to the shaft of the alternator. The moving electrodes are radial copper rods projecting from the

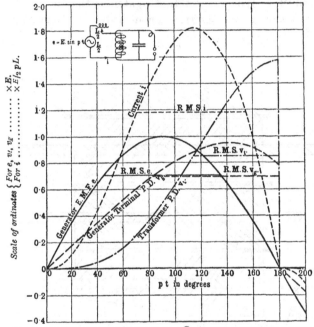

Fig. 29. Low-frequency resonance. One spark per half-period. The v_g curve is for case when inductance of alternator is half total inductance.

edge of a steel disc. The fixed electrodes are the copper discs A, B. These are maintained in slow rotation so that the burnt edge is continually moved on. On loosening nut C, their phase setting can be adjusted and is indicated at the scale D. In high-power transmitters, the Marconi practice is to insert a signalling switch in series with the high-tension winding of the transformer. The switch is controlled through relays by the Morse key, and is provided with powerful air blasts to minimise arcing at the breaks.

While the oscillation is taking place, enormous currents flow back and forth across the spark gap, with consequent large wastage of power there. For example, in the case quoted on p. 39, the amplitude of the spark current at the beginning of the oscillation is 1,000 amperes. The mean resistance of the spark gap during the oscillation might be about ½ ohm. The sooner the energy can be got out of the spark circuit, wherein no useful work is done by the current, into the antenna circuit, wherein the ohmic losses can be kept smaller and whence useful radiation occurs, the more efficient the transmitter. We have seen that tightening the coupling hastens the transference of energy from primary to secondary; but that

Copper Electrodes

Mica Rings

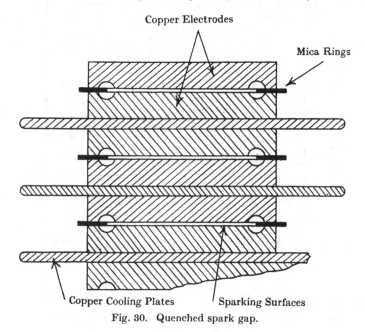

Copper Cooling Plates   Sparking Surfaces

Fig. 30. Quenched spark gap.

retransference from secondary to primary may then occur (Fig. 18). If a tight coupling could be used, and if the spark circuit could be removed or open-circuited immediately it had passed all its energy into the antenna circuit for the first time, we should have the advantage of tight coupling without its defect.

This is precisely what is effected by the use of a special form of spark gap called a "quenched gap." This form of gap appears to have been introduced by E. von Lepel, but its action was investigated and explained first by M. Wien. The ordinary spark gap, between two small metal electrodes separated ¼″ or so in open air, does not become de-ionised quickly enough to prevent re-ignition after the first beat node of current across it. The essential feature of the quenched gap is that the rate of de-ionisation is very much raised. In the form of quenched gap developed by the Telefunken Company, the chief exploiters of quenched spark working, this is achieved by subdividing the gap into a number of very short paths between large well cooled copper or silver surfaces spaced about ·2 mm apart by mica insulating separators, as shown in Fig. 30. Owing to the shape and extent of the sparking surfaces, and their proximity to every portion of the ionised gas, the cooling and

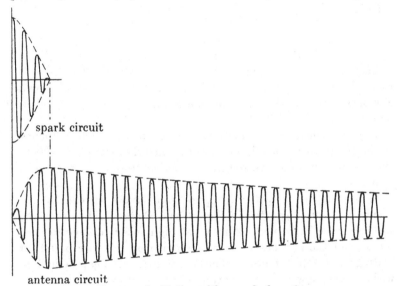

spark circuit

antenna circuit

Fig. 31. Oscillation with quenched spark.

therefore the de-ionising of the gas is extremely rapid. Moreover, enclosing the spark gap from the open atmosphere with only very small volume of air within is found to add to the quenching effect, presumably owing to the sudden rise of gas pressure when the spark occurs. About 1,200 volts per gap is required to produce a spark, and as many gaps are used in series as will bring the total sparking P.D. nearly up to the maximum P.D. available. A complete spark gap of this type is shown in Plate VI. Any desired number of the component gaps can be short-circuited by spring clips inserted between the cooling plates; three of the seven gaps are so short-circuited in the photograph.

With such a quenched gap, it is found practicable to use a coupling coefficient of about 20 per cent. and yet avoid re-ignition after the first beat node. The spark circuit is then extinguished after about $2\frac{1}{2}$ periods, leaving a large proportion of the initial energy in the antenna circuit, which then oscillates with its own independent wavelength and decrement. Fig. 31 shows approximately the trains of oscillation in the two circuits for the case of antenna circuit decrement ·06, a practical value. The current scales in the figure bear no relation to each other; the relative values of the currents must depend, of course, on the relative values of the capacities of the primary and secondary circuits.

Low-frequency resonance is even more important in a transmitter of this type than with the rotary synchronous unquenched spark gap previously described. For with the fixed gap there is no mechanical determination of the instant of sparking, as there is with the rotary gap; and, further, any trace of arcing would obviously prevent the rapid de-ionisation or quenching which is required.

The quenched spark method of generating high-frequency oscillation in an antenna is sometimes termed the method of "shock-excitation" or "impact-excitation": the appropriateness of these terms is obvious.

Various forms of quenched spark gap have been constructed beside the one illustrated in Fig. 30 and Plate VI. The sparking surfaces have been made to move at high speed relative to each other; and the gap has been flooded with oil or other liquid, or hydrogen. In all, the object aimed at is to get the gap de-ionised within a very few periods. Anti-arcing devices are sometimes confused with quenching devices proper, but they should be clearly distinguished. Calculation shows at once that except with the small attenuation factors $\epsilon^{-\frac{R}{2L}t}$ *associated with very great wavelengths, it is not practicable to quench early in the oscillation by the mechanical separation of the electrodes in rotary gaps, or by the violent air blasts sometimes employed to ventilate spark gaps. Thus if the wavelength is 1,000 metres (3×10^5 p.p.s.), and the primary oscillation is to last for three periods as in Fig. 31, the spark must be quenched $\frac{1}{100,000}$ second after it is ignited. The prevention of arcing is another matter. Here we are concerned to sweep out the gap,

* Or $\epsilon^{-\delta \cdot N}$, as on page 25.

as it were, after one oscillation has finished and before another begins; and except with large wavelengths a large proportion of the spark period is available for so doing.

A small non-quenched synchronous spark ship station of $1\frac{1}{2}$ kilowatts is shown in Plate VII. The transmitting apparatus is contained in the cabinet, more clearly exhibited in Plate VIII. A rotary converter (D.C. to A.C., 500 p.p.s.), with spark gap mounted on the shaft, is in the bottom compartment. On the front panel (Plate VII) is seen the remote-control starter, two wheel-handles for adjustment of wavelength and power, and the antenna ammeter. The high-frequency transformer (below), and antenna loading inductance (above) are seen within the cabinet in Plate VIII.

2. CONTINUOUS WAVE METHODS

We have seen that in spark transmitters a train of waves is radiated each time a spark occurs, and that betweenwhiles the power falls appreciably to zero for relatively long periods. Since the amplitude of antenna current decreases according to an exponential function of the form ϵ^{-bt}, it never quite reaches zero; and if the damping is made sufficiently small and the sparks recur sufficiently rapidly, a new oscillation may be set up by a new spark before the oscillation from the preceding spark has died away to a negligible amplitude. If the new spark were made to occur at the right instant, the old and the new oscillations would be in phase, and a continuous oscillation more or less fluctuating in amplitude would be produced*. The antenna current would then approximate to a steady alternating current, the limiting case of a damped oscillation of vanishingly small decrement, and the advantages to be got from "sharpness of tuning," discussed in Chap. III, Sections 2 and 4, would be developed to the maximum. The radiation in space corresponding to a dot or a dash would no longer consist of separated trains of waves, but would be of approximately constant amplitude throughout the dot or dash except during the short periods of growth and dying away immediately following the depression and release of the signalling key. Such

---

* The practical difficulties in utilising this process are very great, and could only be overcome with very long waves. The Marconi Co. have, however, built several high-power, long-wave stations of this type, e.g. at Stavanger (Norway) and Carnarvon. Their system is a development from the synchronous rotary spark gap, and is commonly known as the Marconi "Timed Spark" system. It is a remarkable engineering achievement, but is not likely to be retained in future constructions.

radiation is called "continuous-wave" radiation, commonly abbreviated to "C.W."; and the systems of wireless telegraphy in which the oscillation is sustained throughout the period of the signal are called undamped or C.W. systems, as contrasted with damped or spark systems. A "spark" system, without qualification, always signifies one in which the sparks are far apart so that sensibly independent separated trains of oscillation are produced. Fig. 32 shows the condition during the signalling of a Morse dot, A by a spark transmitter, and B by a C.W. transmitter. It is not true to scale in that the intervals between the separate wave trains in A are usually much longer than the effective length of the train itself; and that therefore, for comparable transmitter powers the amplitude would be much smaller in B than in A.

3. ALTERNATOR METHODS

The combination of damped trains of oscillation produced by sparks is a very roundabout way of generating a sustained alternating current. The alternating E.M.Fs. of low frequencies (25–100 p.p.s.) used in power and lighting circuits are generated by rotating a coil of wire in a fixed magnetic field; or, more usually in large machines, by rotating a magnetic field about a fixed coil of wire. From quite early days, attempts were made to use

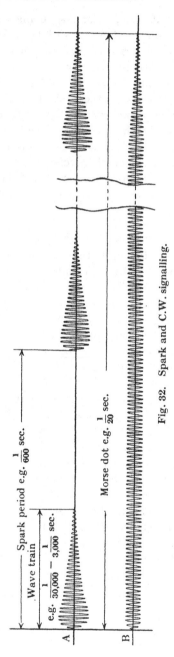

Fig. 32. Spark and C.W. signalling.

the same method for the very high frequencies of wireless telegraphy; but two great classes of practical difficulties arise. The first is mechanical, and is due to the enormous peripheral speeds which are required. The second is the electrical difficulty of keeping within reasonable bounds the eddy-current losses in the iron cores submitted to the rapidly changing fields. Small alternators of the ordinary low-frequency types have been constructed for frequencies up to 1,000 p.p.s. or so; and R. Goldschmidt has even produced machines with wound rotors for frequencies up to 10,000 p.p.s. But for such frequencies the mechanical difficulties are much lessened by a modification known as the inductor type of alternator, in which the rotor carries no winding at all.

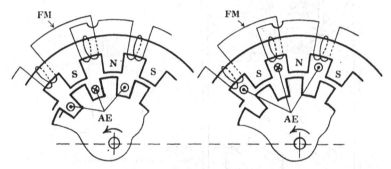

Fig. 33. Inductor alternator.

Fig. 33 shows diagrammatically on the left an ordinary alternator, with stationary field magnet excited by a steady current in the winding FM, and producing an alternating E.M.F. in the rotating winding marked AE. On the right, the same alternator is shown modified by merely transferring the latter winding from the rotor slots to the stator slots. The machine is now an "inductor alternator," and the mechanical difficulties attaching to high rotor speed are greatly reduced. Inductor alternators for frequencies of the order of 500 p.p.s.* are sometimes made after this pattern.

When such an alternator is to generate E.M.F. at frequencies of tens of thousands of periods per second, very small pole pitch, and consequently very small air gap, are necessary; the rotor must be of shape and material to allow enormous speed without excessive stresses from centrifugal forces; special attention must be paid to

* Used for charging the condenser in synchronous spark transmitters—*vide* Section 2 of this chapter.

the reduction of windage losses; and the iron subjected to the fluctuating magnetic field must be as finely laminated, of as high specific resistance, of as small volume, and must carry as low an induction, as possible—conflicting desiderata requiring a nice balance of compromise. Nevertheless high-frequency alternators of this type and of high powers have been constructed in recent years*.

In the high-frequency alternators of E. F.·W. Alexanderson, the arrangement shown in Fig. 34 has been adopted. The

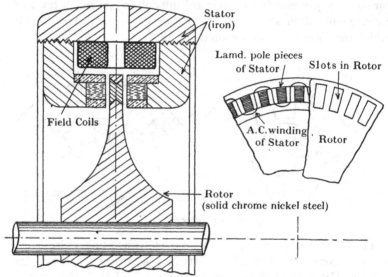

Fig. 34. Alexanderson alternator.

field pole-pieces are very finely laminated; but in the rotor electrical efficiency is sacrified to mechanical strength, and a solid steel disc, with slotted gaps, is used. The gaps are filled in with phospor bronze, riveted over and finished off smooth, in order to avoid the additional windage loss which the irregularities would introduce. In a 2 kW 100,000-cycle machine of which particulars have been published, the rotor is about a foot in diameter, with 300 slots, and is driven at 20,000 revs. per min.; at which speed every ounce of material at the periphery of the rotor demands a radial force of 2 tons weight to hold it in place. The normal air gap is

* For a short account of the principles of various high-frequency alternators, see M. Latour, "Radio frequency alternators," *Proc. Inst. Radio Engineers*, June, 1920.

PLATE VII. COMPLETE SHIP INSTALLATION, 1½ kW (p. 45).
(Radio Communication Co., Ltd.)

PLATE VIII. TRANSMITTER UNIT OF PLATE VII (p. 45).
(Radio Communication Co., Ltd.)

·015 inch. A more recent and much larger machine*, which runs at a speed of 2,170 revs. per min., with an output of 200 kilowatts at 22,000 p.p.s., is shown in Plate IX.

The mechanical difficulties inseparable from the project of an alternator to generate at very high frequency are obviously moderated if it can be arranged that the frequency of the alternator shall be a sub-multiple of the wireless frequency, this initial frequency being stepped up by some form of frequency-multiplier between the alternator and the antenna. This scheme has been realised on a large scale by G. von Arco of the "Telefunken" Company† and Rudolf Goldschmidt‡, using two quite distinct methods of stepping up the frequency.

At Nauen an inductor alternator generated 250 kW§ at 8,000 p.p.s., and between it and the antenna the frequency was doubled twice in succession by a series of static frequency changers of the Joly type, a power of some 100 kW finally being delivered to the antenna at a frequency of 32,000 p.p.s. The frequency-multipliers depend upon the unsymmetrical reaction to changes of positive and negative current of an inductance whose iron core is nearly saturated by an independent steady magnetising current. The principle will be clear on reference to Fig. 35. The steady current I nearly saturates the cores of the two transformers A and B with the magnetic fluxes shown by the dotted lines in the ϕ_A and ϕ_B curves. The two windings supplied with alternating current i_1 of frequency n are connected so that when the magnetomotive forces due to i_1 and I are in the same direction in one core A, they are opposed in the other core B. Owing to the saturation effect in the cores, therefore, the alternating current i_1 produces asymmetric changes of flux above and below the steady current value, the cycle in A being in anti-phase with respect to that in B, as shown by the flux curves ϕ_A, ϕ_B. The E.M.Fs. induced in the third windings are shown separately in the curves e_A, e_B; and their sum is shown in curve $(e_A + e_B)$. This E.M.F. is of double frequency, and is utilised to supply current either to another frequency-doubler or to the antenna, the circuit being tuned so as to offer minimum impedance to the double frequency. The cores and windings of the transformers

* At New Brunswick, U.S.A.
† Gesellschaft für drahtlose Telegraphie, at Nauen (Germany) and Sayville (U.S.A.).
‡ At Eilvese (Germany) and Tuckerton (U.S.A.).
§ More powerful plant has since been installed.

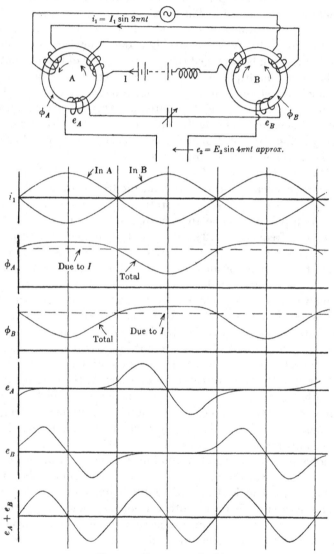

Fig. 35. Frequency-doubler.

are, of course, very specially designed for the high-frequency currents carried; but as there are no moving parts there are no mechanical difficulties, and the configuration of the transformers can be selected with reference solely to the electrical requirements. An objection to the saturated-iron type of frequency-multiplier seems to be the practical impossibility of avoiding a rich mingling of overtones with the fundamental frequency of the output current.

The principle of the Goldschmidt alternator derives from the following theorem, which will be obvious on a little reflection: a stationary alternating magnetic field of amplitude H and frequency n periods per second is identical with a pair of fields of constant strength $\frac{H}{2}$ rotating at $+n$ and $-n$ revolutions per second; consequently if a coil is rotated at n' revs. per second in the stationary alternating magnetic field of frequency n, there are produced in it E.M.Fs. of the frequencies

$$(n' - n) \text{ and } (n' + n).$$

In Fig. 36, S is a stationary coil carrying steady current supplied from the battery. R is a coil rotating in the field of S at n revolutions per second. An n-current* is therefore generated in R. To ascertain what is the resulting E.M.F. in S, reflect that this will be the same as though R were stationary and S were rotating at $-n$ revs. per sec. The stationary R carrying the n-current

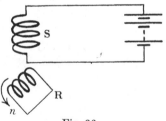

Fig. 36

is equivalent to two rotary fields of speeds $+n$ and $-n$, and hence produces in S two E.M.Fs. of frequencies

$$n - (+n) = 0$$

and

$$n - (-n) = -2n.$$

The resulting $2n$-current in S, now again to be regarded as stationary, is equivalent to two rotary fields of speeds $+2n$ and $-2n$, and hence produces in R, rotating at speed n, two E.M.Fs. of frequencies

$$2n - n = n$$

and

$$2n + n = 3n.$$

* Short for current of frequency n.

And so on. Hence in S is produced a series of currents of frequencies

$$0, \ 2n, \ 4n, \dots$$

and in R a series of frequencies

$$n, \ 3n, \ 5n, \dots.$$

For the currents to become appreciable in magnitude, paths of low impedance must be provided for them. This is indicated in Fig. 37, where provision is made for frequencies up to $4n$, through the circuits containing R or S and the tuning condensers C_n, C_{2n}, etc. C_{4n} is shown as an antenna, wherein the final $4n$-frequency current is utilised. Fig. 37 represents the circuits of a Goldschmidt

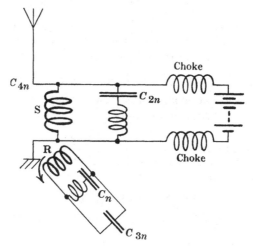

Fig. 37. Goldschmidt alternator.

alternator, S representing the stator winding and R the rotor winding of the machine.

At Tuckerton, the alternator is driven at 4,000 revs. per min.; generates at a fundamental frequency of 10,000 p.p.s.; delivers about 120 kilowatts at 40,000 p.p.s. to the antenna; and has an overall efficiency of about 50 per cent. The core laminations are ·002″ thick; the rotor has 300 poles, is 3 ft. in diameter and weighs 5 tons; and the air gap is less than 1 millimetre.

The alternator has been described first amongst C.W. generators not because it is the most used or in its practical construction the most simple, but because, in the direct-generating form at any rate, it is the simplest to understand. In the case of the Alexanderson machine, which generates directly at the antenna frequency, the alternator winding may be merely connected in series with the antenna (Fig. 38), the single oscillatory circuit thus formed being tuned to the alternator frequency. In actual operation there are of course many practical difficulties. For example, to send Morse dots and dashes, the alternating current in the antenna must be started and stopped under the control of a signalling key; and provision must somehow be made to keep the speed of the alternator constant to within a very small fraction of 1 per cent. whether the key is up or down. Even apart from the question of speed variation, the provision of means whereby a power of hundreds of kilowatts (whether from spark or alternator) may be controlled by a Morse key is an engineering problem of some magnitude. With an alternator as the high-frequency generator, there is the possibility of controlling the output indirectly by controlling the exciting current; and this method is in fact followed in the Goldschmidt system. The distuning

Fig. 38

methods of the "magnetic amplifier," resorted to for the transmission of speech as described in Chapter XI, Section 2, may also be applied in the grosser switching operation of code signalling.

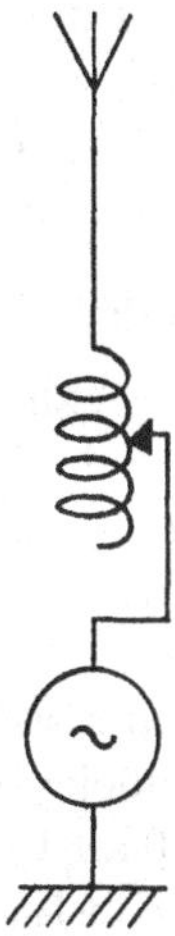

4. ARC METHODS

The use of an arc to produce alternating currents was investigated experimentally and theoretically in 1900 by W. Duddell, who, however, never succeeded in raising the frequency beyond the audible range or in obtaining any but feeble oscillations. V. Poulsen in 1903 modified the arc, and thereby succeeded in producing oscillations not only of wireless frequencies but also of relatively great power. The complete theory of the operation of the Poulsen arc is very complex, and appears even now not fully elucidated. In the physical conditions of the arc there are many independent variables competent to modify profoundly the cycle of operations; and recent researches of P. O. Pedersen appear to show that the hitherto accepted theory, mainly due to H. Barkhausen, while perhaps approximately true for longer arc lengths, does not accord with the observations made with a "normal" Poulsen arc where the gap between the electrodes is very short.

The Duddell arc oscillator is illustrated in Fig. 39, where A is an arc between ordinary pure carbon electrodes in air shunted by a condenser C and an inductance L in series. If the resistance R of the shunt circuit is small, under suitable conditions a steady alternating current is developed in it of frequency approximately $\dfrac{1}{2\pi\sqrt{LC}}$. We will call i_c the instantaneous value of this current, i_a the arc current and I the total current supplied by the battery to the arc and its shunt. We shall assume that the inductance and resistance (if any inserted) between the battery and the arc are sufficiently large to keep the battery current I sensibly constant. At all times, then, the condenser current i_c plus arc current i_a is a constant I.

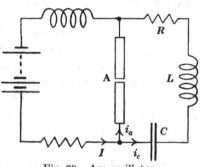

Fig. 39.   Arc oscillator.

The arc is one of that class of conductors for which Ohm's law has no significance except as a definition of resistance. It is a gaseous conductor; or more precisely is a body of non-conducting gas in which the carriers of electricity, the ions or wanderers, are themselves multiplied by the very current they convey. The bigger the current, the more than proportionately better is the conductor; so that the curve relating arc P.D. v_a with arc current i_a is a "falling characteristic" as in Fig. 40. Its slope, $\tan\theta$, is negative and a function of i_a, instead of positive and independent of the current as in non-gaseous metallic conductors obeying Ohm's law.

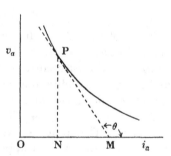

Fig. 40.   Static characteristic.

The characteristic curve can be expressed fairly closely as the hyperbola

$$v_a = A + \frac{B}{i_a}$$

where A and B are constants of the particular arc, independent of

v_a and i_a. The differential equation to the flow of current in the condenser circuit is obtained by summing all the E.M.Fs. in the circuit thus:

$$L\frac{di_c}{dt} + Ri_c + \frac{1}{C}\int i_c dt = v_a = A + \frac{B}{i_a} = A + \frac{B}{I - i_c}$$

$$\therefore\ L\frac{d^2i_c}{dt^2} + R\frac{di_c}{dt} + \frac{1}{C}i_c = \frac{B\dfrac{di_c}{dt}}{(I - i_c)^2}$$

i.e.
$$L\frac{d^2i_c}{dt^2} + \left[R - \frac{B}{(I - i_c)^2}\right]\frac{di_c}{dt} + \frac{1}{C}i_c = 0.$$

Hence as long as the oscillatory current i_c is small compared with the supply current I, the effect of the arc on the oscillatory circuit $CARL$ is to reduce the resistance R by a nearly constant amount $\dfrac{B}{(I - i_c)^2}$, which is the numerical value (say m) of the slope $\tan\theta$ of the falling characteristic (Fig. 40) around the particular arc current ON at which the arc is burning.

If, therefore, the $CARL$ circuit were disturbed in any way and then left, it would oscillate approximately thus:

$$i_c = \epsilon^{-\frac{(R - m)}{2L}t}\sin\left(\frac{1}{\sqrt{LC}}t - \phi\right).$$

If $m < R$, this is a decrescent oscillation such as we have already encountered in spark circuits (Fig. 15). But if $m > R$, the exponential term increases with time. The tiniest fluctuation in the $CARL$ circuit would then set it oscillating, at approximately its natural frequency determined by L and C, with an amplitude which would continue to grow until either our mathematical approximation $i_c << I$ were no longer applicable or until some new physical conditions supervened.

The curve of Fig. 40 is the arc characteristic for steady current. It obviously can not hold for rapidly varying currents; for ions when formed take time to disappear by convection or otherwise, and moreover the thermal condition of the arc at any instant must depend largely on the currents at preceding instants. At any value of the current i_a when it is decreasing, the temperature will be higher, and the P.D. v_a therefore lower, than the static values corresponding to that current. The kinetic P.D. will thus be higher than the static P.D. when the current is increasing, and lower when it is decreasing. The kinetic characteristic of the Duddell oscillating arc therefore assumes some such form as Fig. 41; and it is obvious that the frequency of the oscillation must profoundly influence the behaviour of the arc. With an indefinitely great frequency there could be no negative slope at all, for there would be no time for the amount of ionisation to follow the changes of current.

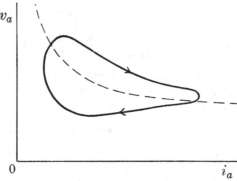

Fig. 41.   Kinetic characteristic.

For some such reasons as these, neither the power nor the frequency of oscillation from the Duddell arc can be made large. The kind of values in Duddell's experiments were:

$$L = 5mH \qquad C = 1 \text{ to } 5\mu F \qquad R = \cdot 4\Omega$$

$$\text{R.M.S. current} = 3A \qquad \text{Frequency} = 2,000\text{--}10,000 \text{ p.p.s.}$$

The Duddell oscillations are characterised by the fact that the arc is never extinguished, the maximum value of the oscillatory current i_c being less than the steady feed current I, as in Fig. 42.   Such

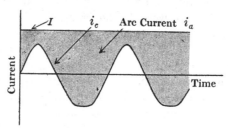

Fig. 42.   Duddell oscillation.

arc oscillation is called "Oscillation of the first type," or here for short "α-oscillation."

The arc became practicable as a generator of high-frequency currents for wireless telegraphy in the hands of V. Poulsen in 1903. The modifications made by Poulsen were threefold, and a Poulsen arc is illustrated diagrammatically in Fig. 43. These modifications are:

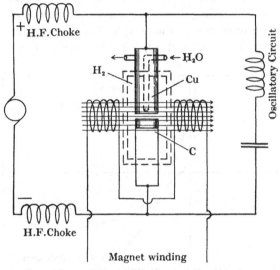

Fig. 43. Poulsen arc.

(i) The arc is enclosed in an atmosphere of hydrogen or coal gas or alcohol, kept as cool as possible by water jacket.
(ii) The anode is of copper, cooled by internal water circulation.
(iii) A powerful magnetic field is maintained across the arc, acting as a magnetic blast driving the arc away from the electrodes.

Fig. 44 shows the marked effect of (i) and (ii) even on the static characteristic of the arc; and it is to be expected that the effect on the kinetic characteristics at high frequencies will be still more pronounced. The hydrogen atmosphere favours rapid de-ionisation, hydrogen of all gases possessing the highest thermal conductivity and the highest diffusion coefficient.

It is clear that α-oscillation can be much more easily produced with the steeper characteristic of Fig. 44, the limiting value of resistance of the oscillating circuit being proportional, as we have seen, to the steepness of the curve. An oscillating arc must always start up with the α-oscillation; but with the Poulsen arc, the oscillation once started builds up into what is called "oscillation of the second type," or here for short "β-oscillation," characterised by the oscillatory current having so large an amplitude that during part of the cycle the arc is actually extinguished. It is re-ignited later in the cycle at the moment when the P.D. between copper and carbon reaches the ignition or sparking value, which greatly exceeds the burning value—the more so the more completely de-ionisation has been effected during the period of extinction.

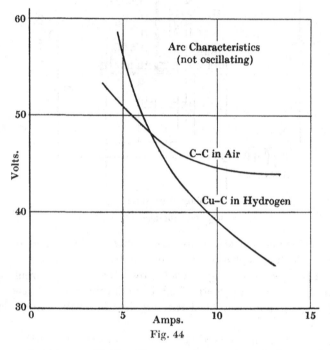

Fig. 44

Whereas in the α-oscillation there can be no abrupt changes in the current or P.D., since the representative-point on the arc characteristic never leaves the smooth curve, in the β-oscillation abrupt changes do occur.

In Fig. 45 an attempt is made to trace the cycle of current and P.D. changes with β-oscillation. The complete cycle is divided into two parts: an epoch T_1 during which the arc is burning, and an epoch T_2 during which it is extinct. In the normal Poulsen working T_2 is very much shorter than T_1. Let us examine in succession (in Fig. 45) the course of the condenser current i_c and the arc current

i_a, and the condenser P.D. v_c and the arc P.D. v_a; bearing in mind always that the chokes in the supply mains are supposed to make I sensibly constant, a condition approximately realised in practice.

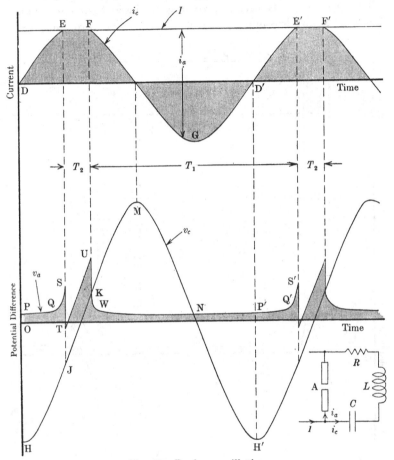

Fig. 45. Poulsen oscillation.

Condenser current DEFGD' (*area shaded*). At D the condenser is fully charged in the negative direction, and along DE receives a current growing approximately sinoidally. At E this current equals the total invariable supply current I, so that no current is available for the arc, which is therefore extinguished. Along EF the condenser receives the whole of the supply current I, and inevitably charges up more and more until its P.D. reaches such a value as at F to re-ignite the arc. This provides again a by-path to shunt away from the condenser some of the steady I. The oscillatory circuit *CARL* now being again closed through the arc, the condenser C discharges along FG with approximately sinoidal current, the arc tending to have less and less effect as the current through it rises.

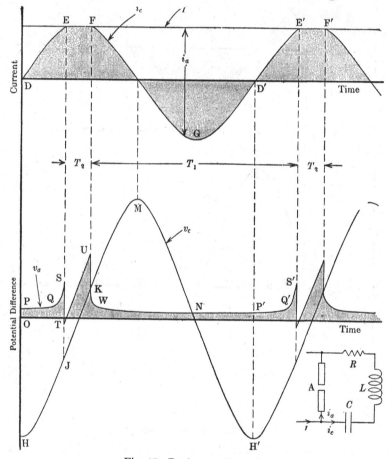

Fig. 45. Poulsen oscillation.

Except near the ends F and E′, the condenser discharge FGE′ is not very different from what it would be if an ordinary metallic conductor replaced the arc.

Arc current, i_a. This is the difference between the horizontal I line and the curve DEFGD′. i_a is always positive.

Condenser P.D., HJKMNH'. This is a very nearly sinoidal curve with its maxima opposite D and D', the portion JK being slightly changed from the sine shape owing to the horizontality of EF. If DEFGD' had been truly sinoidal, HJKMNH' would of course have been precisely another sinoidal curve lagging by $\frac{1}{4}$ period.

Arc P.D., PQSTUKWP' (*area shaded*). At P the arc current is large, so that the P.D., OP, is small. In the region Q the arc current is becoming small, so that the P.D. rises rather suddenly to the extinction voltage at S. Since i_c is now steady ($= I$), v_a must be the P.D. of the condenser plus the ohmic drop in R; i.e. under EF, $v_a = v_c + RI$. v_a therefore rises along TU at a constant height TJ $=$ UK $= RI$ above JK to the striking voltage at U. U is generally higher than S on account of the progress in cooling and de-ionising during the extinct period T_2. On re-ignition at U and during the subsequent growth of i_a, v_a drops rapidly at U, passes through a sharply curving region at W corresponding to that at Q, gradually drops to a minimum height under G, and slowly rises again to P'.

What now of the power thrust into RLC, where energy is stored (in L and C) and lost or utilised (in R)? The arc P.D. v_a is the E.M.F. impressed on this circuit RLC, and the rate at which at any instant energy is passing into it is $v_a \times i_c$, i.e. the product of the ordinates of the curves whose areas are shaded. During the middle portion of T_1 while i_c is negative, i.e. while the arc current exceeds the supply current I, this power is negative and proportional to the large-current P.D. of the arc. During all the rest of the cycle, except perhaps for a brief moment just after extinction at T, the power into RLC is positive, and plainly depends largely upon the heights S and U, i.e. upon the extinction and ignition voltages.

The following qualitative conclusions can be drawn from the consideration of these curves:

For large power delivered to the oscillatory circuit,

(*a*) I should be large;

(*b*) extinction voltage large;

(*c*) ignition voltage large;

(*d*) resistance of oscillatory circuit large.

For small power taken back from the oscillatory circuit,

(*e*) I small;

(*f*) large-current P.D. of arc small.

In the conflict between (*a*) and (*e*), (*a*) must obviously preponderate. The maximum current in the arc is about $2I$, and the size of the electrodes must be made sufficient for this current. For (b), we need a long arc, well de-ionised; and for (c) the same; but it must be noticed that whereas the ignition arc length is the shortest distance between the electrodes, the extinction length is much greater owing to the action of the magnetic field ("magnetic blast") which both bows the arc and separates the craters during the burning epoch T_1. The hydrogenous atmosphere and the coolness of the anode favour de-ionisation. (d) is under control, as the oscillatory circuit can, within limits, be designed to suit the arc. For (f) a short arc is required; this conflicts with (c), but owing to the magnetic blast not with (b).

These considerations confer some insight into the rationale of Poulsen's modifications of the Duddell arc, and into the conditions governing the design of Poulsen arcs*.

Nothing has been said about steadiness of operation, which is particularly important as regards constancy of wavelength; nor about freedom from overtones of the fundamental frequency which may cause great disturbance to neighbouring receiving stations. These are practical matters of great weight, and the future of the arc is likely to depend as much on its perfection in these respects as in any others. It appears that steadiness is largely influenced by the strength of the magnetic field. The correct strength depends on the frequency of oscillation, amongst other factors, so that the magnet excitation requires adjustment when the wavelength is changed. To provide against gross variations in the arc as the carbon is dissipated, the cathode (carbon) is kept in slow axial rotation by a small electric motor.

The oscillatory circuit across the arc generally consists of the antenna itself, as in Fig. 46. And since the arc must be left burning whether the Morse key is up or down, signalling is effected either by switching in a dummy (non-radiating) antenna in place of the real one, or else by slightly changing the wavelength under the action of the signalling key, as in Fig. 46. The latter is the usual method at the present time. It is open to the objection that power is being consumed just as much during spacing as during marking. Not only is the spacing power sheer waste from the transmitting

---

* For a more thorough and quantitative investigation, see P. O. Pedersen, "On the Poulsen arc and its theory," *Proc. Inst. Radio Engineers*, Vol. v, No. 4 and Vol. vii, No. 3. A complete bibliography of the subject is included.

standpoint, but it is a potential nuisance to other receiving stations. The existence of the spacing radiation implies the allocation of two wavelengths to the one station—an objection of ever increasing weight as the world becomes more congested with high-power stations. The main supply current itself is commonly used to energise the field magnets producing the magnetic blast. The arc is started up by "striking" as in ordinary arc lamps. A condenser, shown dotted in Fig. 46, is sometimes shunted across the arc, so that only a portion of the oscillatory current flows through the arc.

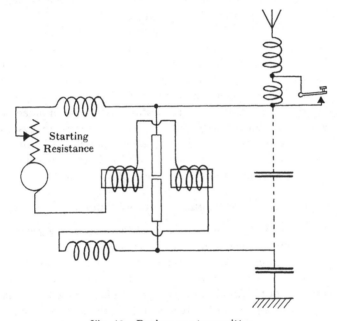

Fig. 46. Poulsen arc transmitter.

An idea of practical dimensions may be had from the following figures for a large arc transmitter. Suppose the D.C. generator supplies 200 amps. at a pressure of 500–800 volts (depending upon the antenna resistance). The R.M.S. antenna current would be very approximately $\frac{200}{\sqrt{2}} = 141$ amps. The antenna might have a capacity of ·05–·10 microfarad, a natural wavelength (without loading inductance) of (say) 2,000 metres, and be loaded up to radiate at (say) 6,000 metres. The efficiency of the power conversion from

D.C. to high frequency would be in the neighbourhood of 40 per cent. The actual signalling key would not be connected with the antenna directly, as in Fig. 46, but would control some form of multi-contact signalling switch through intermediary relays. The difference between marking and spacing wavelengths might be 1 per cent. or 2 per cent.

Plate X shows an Elwell-Poulsen arc suitable for an input of 10–15 kilowatts. Plate XI shows one of the largest arcs yet put into use, constructed by the Federal Telegraph Co., for 500 kilowatts input. Plate XII shows the antenna loading coil and signalling switches for use with a 30-kilowatt arc.

Until recently the Poulsen arc was the only practical means of producing C.W. radiation, but was nevertheless surprisingly little used. To-day, seventeen years after its invention, and despite the subsequent introduction of various other types of high-power C.W. generator, it has become of the greatest importance for long range work. Indeed it can hardly be contested that at the moment the arc stands pre-eminent amongst wireless transmitting systems, whether in respect of the amount of power put into the antenna at individual stations, or in respect of the total power consumed in all stations of any one type of over (say) 50 kilowatts*.

5. THERMIONIC TRIODE METHODS

The newest of the four great classes of high-frequency generator, developed mainly during the war, is that of the three-electrode thermionic vacuum tube, or triode. There can be little doubt that in a very few years practically all low-power wireless transmitters will be of this kind; and if the great efforts now being made to obtain very large outputs—tens or hundreds instead of units of kilowatts—are as successful as may be expected, the alternator and the arc, as well as the spark, will probably wholly give place to the triode†.

Consideration of the triode as generator is postponed to Chapter IX.

* A map showing the large Poulsen arc stations of the world is given by C. F. Elwell in "The Poulsen system of radiotelegraphy," *Electrician*, May 28, 1920.

† A critical review of the several types of wireless transmitter available, at the time of the report, for long-distance wireless communications, is contained in the Report of the "Imperia. Wireless Telegraphy Committee, 1919–20," published by H.M. Stationery Office. For the imperial communications under consideration, the committee recommended the use of triode transmitters.

PLATE IX. ALEXANDERSON ALTERNATOR, 200 KW, 22,000 p.p.s. (p. 49).

(The Wireless Press, Limited.)

SOLENOID.
ALCOHOL FILLING DISH.
RESERVOIR.
WATER OUTLET OF LID.
WATER INLET OF CHAMBER.
WATER INLET OF ANODE.
POLE PIECE ADJUSTING SCREW.
WATER OUTLET OF ANODE.
TERMINAL BLOCK.
FIELD COIL.
BASE OF MAGNETIC CIRCUIT.

VALVE.

LID BOLT.
LID.
CHAMBER.
GASKET.
PORCELAIN.
CATHODE PLATE.

ARC STRIKING & ADJUSTING MECHANISM.
CARBON HOLDER.
CATHODE SHEATH.
CATHODE WATER INLET.
INSULATING TUBE.
WEDGE.

PLATE X. ELWELL-POULSEN ARC, 10–15 κW (p. 63).
(C. F. Elwell.)

CHAPTER V

THE DETECTION OF HIGH-FREQUENCY CURRENTS

1. The main methods

WE have seen how currents of high frequency may be produced in a transmitting antenna, and that these currents set up a condition in surrounding space such that minute electromotive forces, proportional to these currents and of the same frequency, make their appearance in receiving antennae hundreds or thousands of miles away over Earth's surface. The current in the transmitting antenna being switched on and off, or modified in frequency by a signalling key manipulated by a telegraphic operator, our system of communication is complete if we provide means whereby another operator, at the receiving station, can perceive the tiny high-frequency currents in the receiving antenna. The necessity for unusual devices to accomplish this end is due to two independent facts: (i) the E.M.Fs. produced in the receiving antenna are extremely small (of the order of a microvolt), and the power available for affecting a detector is extremely small (of the order of a micro-microwatt); and (ii) the frequency of these E.M.Fs. is extremely high (say between a tenth and ten times half a million periods per second).

Two main methods present themselves for detecting high-frequency currents. We may observe, indirectly, some temperature effect of the high-frequency current (as in the ordinary hot-wire ammeter); or we may more or less completely rectify the alternating current into a current whose mean value is no longer zero, and then use some polarised electromagnetic indicator (such as a telephone, or d'Arsonval galvanometer)*. Instruments of the hot-wire ammeter type are used for measuring the large high-frequency

---

* There are other methods, notably: the coherer (obsolete), and its modern substitute, the author's Oscillatory Valve Relay, which are trigger devices requiring independent restoration after they have been actuated; the Marconi Magnetic Detector (obsolescent), in which the magnetic retentivity of iron is, as it were, shaken loose by the high-frequency current; and the electrodynamometer telephone (obsolete).

currents at the transmitter, but are too insensitive for anything small. Thermo-couples heated by a fine wire carrying the high-frequency current, and associated with some form of d'Arsonval instrument, are much more sensitive, and are used for measuring small high-frequency currents in the laboratory. If the thermo-couple with its heater is enclosed in a vacuous bulb, it becomes both independent of draughts and much more sensitive. Plate XIII shows a very convenient portable combination of vacuo-thermo-junction and millivoltmeter, the connections of which are given in Fig. 47*.

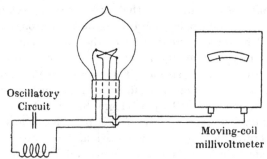

Oscillatory Circuit

Moving-coil millivoltmeter

Fig. 47. Measurement of H.F. current by thermo-couple.

But the rectifier method is capable of being made so very much more sensitive, and with such marked simplicity—particularly when the high-frequency current is to be detected rather than measured, so that a telephone receiver can be used as the indicating instrument and the ear as the perceiving organ—that almost without exception present-day wireless telegraph and telephone receivers depend upon some form of rectifier. The term "detector" should properly connote the instrument whereby the high-frequency currents are made to produce an effect perceptible to the operator, and the rectifier is only one part of one form of the complete detector; but it has come about that detector and rectifier are often used as almost synonymous terms.

* With such a combination full-scale deflection may be got with a power in the heater wire of about 500 microwatts. Much greater sensitivity can, of course, be obtained by substituting a suspended reflecting galvanometer for the portable pivoted instrument.

2. The crystal detector

It is the case with certain minerals, usually crystalline, that the resistance to the flow of electricity across the contact between a piece of metal and a piece of the mineral, or between two pieces of different minerals, depends largely on the value of the current, even when the current is extremely small. That is to say, as in the case of the arc, Ohm's law has no significance except as a definition of the term resistance, the characteristic between P.D. and current having pronounced curvature. Now any conductor with curved characteristic acts as a rectifier, in the sense that if there is applied to it an alternating E.M.F. whose mean value is zero, the mean value of the resulting current is not zero*. This mean value is spoken of as the "rectified current": it is the current which would be indicated by a moving-coil microammeter in series with the rectifier.

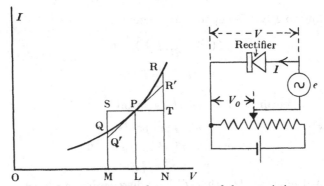

Fig. 48. Rectification by curvature of characteristic.

Let a steady P.D. $V_0 = $ OL (Fig. 48) be applied to the rectifier, whose characteristic is the curve QPR, and produce through it a steady current $I_0 = $ LP; and let a small sinoidal E.M.F. of amplitude $e = $ LM $= $ LN be superposed in the detector circuit to represent the incoming high-frequency signal. If the characteristic had been the tangent Q'PR' instead of the actual curve QPR, e would have produced a sinoidal current of amplitude SQ' $= $ TR' $= m_0 e$, where m_0 is the slope of Q'R'. But owing to the curvature of the characteristic, the alternating current through the

* Except where the mean position of the representative point on the characteristic is at a point of inflection with curvatures symmetrical on each side.

detector has a negative maximum SQ smaller than the positive maximum TR, and the mean change of current produced by e is not zero but has a positive value. If the curve had been convex upwards instead of concave, the signal e would have produced a decrease instead of an increase in the mean current. It is obvious that for large rectifying effect, rapid change of slope of characteristic is required.

It is important to see how the value of the rectified current depends on the amplitude of the signal É.M.F. Let the characteristic curve be expressed by the equation

$$I = f(V).$$

Then if i is the change of current (instantaneous value) due to a small change of P.D. v

$$(I + i) = f(V + v)$$
$$= f(V) + vf'(V) + \frac{v^2}{2}f''(V) + \frac{v^3}{3}f'''(V) + \dots$$

according to Taylor's expansion, where

$$f'(V) \equiv \frac{d}{dV}f(V), \quad \text{etc.}$$

$$\therefore \ i = vf'(V) + \frac{v^2}{2}f''(V) + \dots$$

$$\therefore \ \text{mean } i = \text{mean } vf'(V) + \text{mean } \frac{v^2}{2}f''(V) + \dots.$$

Now
$$v = e \sin pt$$
$$\therefore \ \text{mean } v, v^3, v^5 \dots \text{ each } = 0$$

and
$$\text{mean } v^2 = \tfrac{1}{2}e^2.$$

So that if higher orders than f''' are disregarded, we have

$$\text{Rectified current} \equiv \text{mean } i$$
$$= \text{mean } v^2 \times \frac{1}{2}\frac{d^2I}{dV^2}$$
$$= \frac{e^2}{4}\frac{d^2I}{dV^2}.$$

The important point to notice here is that over any small region of the curve determined by V_0, the rectified current is proportional to the *square* of the signal E.M.F.; so that the rectifier becomes less and less efficient as the signal becomes weaker. Thus if the current in the sending antenna is halved, the E.M.F. and current in the receiving antenna are halved, the high-frequency P.D. across the crystal detector is halved, and the rectified current is quartered. Expressed in terms of power, the sending power is divided by four, and the "rectified" power (perceived by the receiving operator) is divided by sixteen. This progressive failure of the detector as the received signals are weakened is a vital factor in estimating the possibilities of various types of wireless receiver; it shows the importance, where a rectifier is used in this simple way, of amplifying a weak signal before rather than after rectification*.

The instrument most commonly used for indicating the rectified current to the operator is a telephone receiver (of the ordinary permanent-magnet Bell's type), with two ear-pieces clamped over the operator's head. A complete receiving circuit, embodying a crystal detector as in Fig. 48, is shown in Fig. 49. The telephone is shunted by the condenser C to provide a path of small impedance for the high-frequency currents which traverse the detector. Owing to the asymmetric conduction through the detector, this condenser receives unequal positive and negative charges during the two half-cycles, and

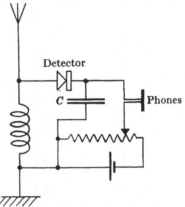

Fig. 49. Receiving circuit with crystal detector.

the net charge thereby accumulated in it subsequently passes through the telephone. The operation will be best understood from a diagram, showing the signals received from a spark transmitter while the transmitting key is held depressed.

Fig. 50 depicts the events between the beginning of one spark (at the transmitter) and the beginning of the next. Curve A shows the train of oscillation in the receiving antenna. Curve B represents the gradual rise of condenser P.D. as the rectified current passes into it, and the subsequent fall as the charge passes out through the telephone. Curve C shows the mean value (with respect to the high frequency) of the change of current through the detector due to the signal (the steady non-signal current I_0 not being shown). And finally curve D shows the resulting change of current in the

* See further under Heterodyne Reception (with triodes), Chapter VIII, Section 4.

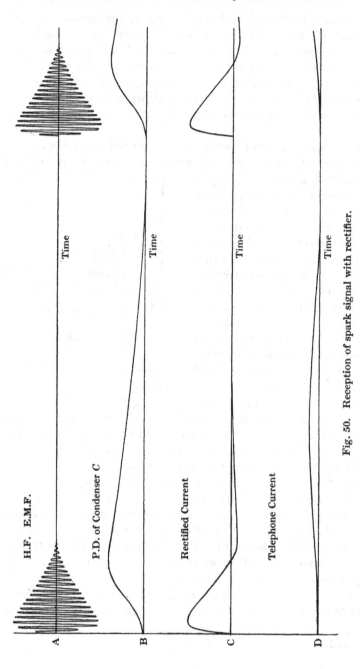

Fig. 50.   Reception of spark signal with rectifier.

telephone. The curves are not drawn to scale, but indicate, with the necessary exaggeration, the phase relations subsisting between the quantities named. Owing to the back P.D. of the condenser, rectification ceases (i.e. the rectified current becomes zero) just before the end of the oscillation, and a small reverse current thereafter flows back through the detector from the condenser until the latter is quite discharged. The area of curve D is equal to the difference between the positive and negative areas of curve C. The diaphragm of the telephone is attracted (or repelled) by the condenser discharge of curve D; so that the operator's ear-drum receives one little blow for every spark at the transmitter, and he hears a musical note whose pitch is the spark frequency. If the sparking is not regular (as in the spark transmitter of Fig. 27), he hears a sound of correspondingly scratching or buzzing character.

The condenser across the telephones (C, Fig. 49) is sometimes omitted in diagrams, and even in real circuits. But it is only the accidental self-capacity of the telephone winding and connections which permits this, some shunting capacity across the telephones being absolutely necessary. The detector current due to the signal is in the form of very brief impulses repeated at wireless frequency, and if the telephone winding were free from capacity, its reactance to current of such frequency would be of the order of megohms. Although condenser C plays no appreciable part in the tuning of the oscillatory circuit, it is sometimes made adjustable for the purpose of giving some rough acoustic (tone-frequency) resonance in the telephone circuit; but the electrical dimensions of an ordinary telephone are such that electrical resonance is not very marked. The telephone-condenser circuit is, however, not actually aperiodic; so that curve D in Fig. 50 should accurately be shown as descending to negative values of current before finally becoming zero.

A great variety of substances have been used to give the rectifying contact in crystal detectors. Two very common combinations are carborundum (SiC) and steel, and zincite (ZnO) and chalcopyrite ($CuFeS_2$). A fragment of the crystal is mounted in a small brass cup with solder or fusible alloy, and the other member of the pair is pressed against it by means of some spring adjustment. Plate XIV shows a simple pattern, in which a hard polished steel disc is pressed on to the point of a spur of selected carborundum. A good feature of the carborundum detector is that, presumably owing to its hardness, considerable pressure may be applied. Representative characteristics are given in Fig. 51. The curve for the carborundum detector shows that if it is to be sensitive to weak signals, a steady polarising P.D. of ·6 volt or more must be applied. The best value varies widely with different samples of crystal, and may even be reversed in sign. The potential divider shown in Figs. 48 and 49 enables the operator to adjust V_0 by trial to the particular value giving greatest sensitivity. In the case of the zincite-chalcopyrite

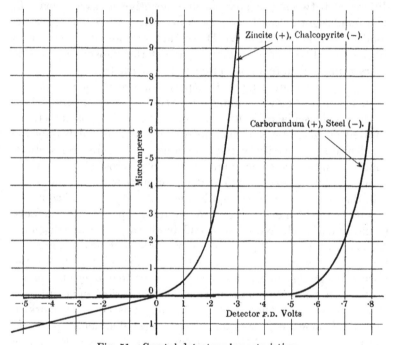

Fig. 51.   Crystal detector characteristics.

detector, although a polarising P.D. of ·2–·3 volt would improve sensitivity with very weak signals, there is fair curvature quite close to zero P.D.; so that with this detector, and others like it, the battery and potential divider are sometimes omitted for the sake of simplicity.

An objection to all crystal detectors is that their sensitiveness is liable to be destroyed by extra powerful signals, e.g. atmospherics; in which event the contact has to be reopened and a new sensitive setting found. Carborundum is less troublesome in this respect than most other crystal detectors.

What determines the sharply curved characteristics exhibited by the materials used in crystal detectors is not fully understood. Moreover it is certain that, at least in some detectors, the static characteristic (e.g. Fig. 51) is not accurately followed when the P.D. changes rapidly. The fact, as W. H. Eccles points out, that the departure increases with the rapidity of change, indicates that thermoelectric effects play an important part in the rectifying action. On the other hand it is certain that in many detectors the effect is not due to the mere thermo-couple E.M.F.; and it seems probable that in different contact detectors rectification depends on different physical properties of the two materials*.

Telephones used as receivers for the mere musical notes of wireless telegraphy usually have rather thinner diaphragms than those required for articulation of the spoken word. Mechanical resonance is not very sharply marked, but, such as it is, occurs usually in the neighbourhood of 1,000 periods per second. Inspection of the curves of Fig. 51 shows that in the sensitive regions the slope-resistances $\frac{dV}{dI}$ of the detectors are in the neighbourhood of 100,000 ohms. The telephones fed from the detector should have a correspondingly high impedance. They are therefore wound with extremely fine wire, resulting usually in a total resistance of 2,000–8,000 ohms. To avoid the use of the very fine wire, and to save the telephone from possible damage caused by the steady non-signal current†, a step-down transformer is often interposed, and the telephones are then wound to some more convenient resistance such as 100 ohms.

3. THERMIONIC DETECTORS

The foregoing examination of rectification from the simple consideration of the static characteristic may be an incomplete account of the behaviour of many crystal detectors; but it does closely apply to rectifiers of the thermionic class in which the physics is much more determinate. These are dealt with in Chapter VIII.

---

* See D. Owen, "The laws of variation of resistance with voltage at a rectifying contact of two solid conductors, with applications to the electric wave detector," *Proc. Phys. Soc.* of London, June, 1916. Abstract in *Electrician*, Sept. 1916.

† More especially when the crystal is replaced by thermionic detectors, hereafter described, with which the steady current may be several milliamperes.

4. CONTINUOUS WAVE RECEPTION

With spark transmitters, every spark produces a train of oscillation, with its accompanying change of current in the receiving telephones; and between the sparks the rectified current falls sensibly to zero. A sound is thus produced in the telephone as long as the transmitting key is held depressed. With C.W. transmitters, on the other hand (see Fig. 32), the rectified current would rise at the beginning of the Morse dot or dash and would remain substantially constant until the end; so that the telephone would indicate the occurrence of the signal only by a single faint click at the beginning and at the end. No summation of the received signal, psychological, electrical or mechanical, would take place, and the telephone would no longer be a suitable receiving instrument.

In order to retain aural reception in C.W. telegraphy, therefore, some means must be provided for changing the steady rectified current into an unsteady one, varying preferably at a musical rate. This may obviously be done by the use of an independently operated interrupter or switch connected somewhere in the receiving circuits. It was done in a better way in the earlier days of the Poulsen arc by P. O. Pedersen's "Tikker," a form of vibrating or crazy-contact switch which combined the functions of interrupter and detector. These methods have, however, entirely given place to the "heterodyne" method, first proposed by R. A. Fessenden in 1902, but only brought into common use with the advent of the triode oscillator.

The principle of the heterodyne is the excitation of the receiving circuit by a steady locally produced oscillation of frequency differing slightly from that of the incoming signal. In the absence of a signal, this produces a steady rectified current imperceptible in the telephones. During a signal, the two oscillations are combined, with the interference or beat effect familiar in acoustics when two musical tones of slightly differing pitch are mingled. Suppose the E.M.F. in the detector circuit due to the local oscillation is $a \sin pt$; and that due to the incoming signal, when present, is $b \sin qt$. The resulting E.M.F. is

$$a \sin pt + b \sin qt = b \left(\sin pt + \sin qt\right) + (a - b) \sin pt$$

$$= 2b \cos \frac{p - q}{2} t . \sin \frac{p + q}{2} t + (a - b) \sin pt.$$

The second of these terms produces a steady, and therefore sound-less, rectified current; the first is an E.M.F. whose frequency is the mean of that of the two oscillations $\left(\text{given by the factor} \sin \dfrac{p+q}{2}\right) t$, and whose amplitude becomes twice that of the signal (viz. $2b$) whenever the factor $\cos \dfrac{p-q}{2} t$ becomes numerically equal to unity, viz. $\dfrac{p-q}{2\pi} = (n_1 - n_2)$ times per second, where n_1 and n_2 are the frequencies of the two oscillations. The superposition of the two oscillations is portrayed graphically in Fig. 80.

The local oscillation has therefore converted the steady (and therefore soundless) amplitude b of the signal into an amplitude varying between 0 and $2b$ at a frequency $(n_1 - n_2)$. By adjusting the local oscillator so that n_1 differs suitably from n_2, we make a C.W. signal declare itself by an audible tone in the telephone. For example, suppose the wavelength of the incoming signal is 3,000 metres; $n_2 = 100,000$ p.p.s. If we distune the local oscillator by (say) 1 per cent., making $n_1 = 101,000$ or 99,000 p.p.s., the signal in the telephones is a shrill pure tone of a pitch (1,000 p.p.s.) to which the telephone-ear combination is very sensitive, and which is quite distinct from the rustling or crackling sounds due to atmospherics or other causes.

A comparison between continuous wave and spark telegraphy, as judged at the receiving station, may be summarised as follows:

Receiving apparatus. The same with both, except that with C.W. a local oscillation must be superimposed on the receiving circuits.

Sharpness of tuning. With C.W., if the receiving circuits are suitably proportioned, the sharpness of the resonance relation between wavelength of signal and tuning adjustment of the receiving circuits is much more pronounced. With spark, since the incoming signal itself has a large decrement, it is futile to strive after improvement by extreme reduction of receiver de-crements.

Telephone note. Spark has the theoretical advantage that the identity of the sending station is defined by wavelength *and* note, whereas with C.W. the note is determined at the receiver. In practice, however, the purity of the C.W. heterodyne note, and the operator's ability to adjust it to any pitch which suits his

apparatus or which aids him in dodging the particular interference he is experiencing at the moment, quite outweigh the
advantage of the spark's characteristic note.

Sensitiveness of receiver. With equal distance and power of
transmitter, the C.W. receiver is very much superior to the spark.
This is mainly due to an improvement in rectifier efficiency caused
by the local oscillation, as explained in Chapter VIII, Section 4.

Further consideration of heterodyne reception is conveniently
postponed till Chapter VIII, Section 4, where the behaviour of the
rectifier is examined quantitatively for the case where the characteristic consists of a horizontal portion and a straight inclined
portion connected by a curved portion.

5. RECEIVING CIRCUITS

In Fig. 52 A, the power delivered by the antenna to the detector
circuit is obviously affected by the position of the tapping point
on the inductance coil in the antenna circuit; and the larger the
resistance of the detector the higher that point should be. The
higher the tapping point, the larger the resistance introduced into
the antenna by the connection of the detector; and it is easy to show
that for a steady C.W. signal the power delivered to the detector
is a maximum when the total antenna resistance with detector is
twice its resistance without detector.

It often happens that it is not convenient to have so much
inductance inserted in the antenna that a suitable tapping point
can be arrived at. The arrangement of Fig. 52 B or C is then
resorted to. In B, a step-up transformer effect is sought with a
tightly coupled aperiodic secondary coil, that is a coil whose
natural frequency is far below resonance, so that tuning of the
secondary circuit is unnecessary. In C, a tuned secondary circuit
is provided, with variable coupling between primary and secondary.
If C_2 and L_2 are selected so that the decrement of the C_2L_2 circuit
with detector is very small, this arrangement possesses the advantage that a light coupling gives sharper resonance (to a lightly
damped or C.W. signal) than would be given by arrangements A or B,
where the only oscillatory circuit is the antenna circuit of relatively
high decrement.

The manner of connecting the detector to the antenna or
associated oscillatory circuit can obviously be varied in many
ways; but with all circuits the action can be similarly analysed in

terms of the damping of the antenna circuit, and of any associated oscillatory circuits, by the detector. In addition to variable couplings, receivers are sometimes provided with switching arrangements by which the type of connection between detector and antenna can be varied; e.g. from A or B (Fig. 52) while searching for a signal of perhaps unknown wavelength, to C when communication has been established and sharp tuning with sub-maximum signal strength becomes desirable in order to minimise interference from other stations.

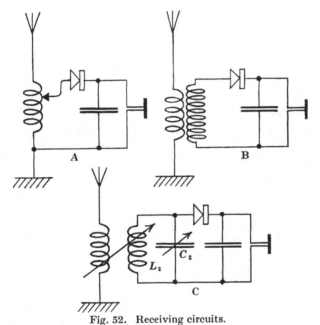

Fig. 52. Receiving circuits.

Receiving circuits, since they carry only tiny currents of low voltage, can be made very compact, particularly for the smaller wavelengths. They are often boxed up to include inductance coils, switches, adjustable coupling devices, potentiometers, variable condensers, etc. in divers combinations for different purposes. An example is shown in Plate XV, where three receiving inductance coils are clearly visible, two of them coupled together by variable mutual inductance. Plate XVI shows the usual type of continuously variable condenser with air dielectric for tuning purposes. A complete receiver is seen on the table in Plate VI.

CHAPTER VI

THE THERMIONIC TUBE

1. Its practical importance

WITHIN the last six years, the art of wireless telegraphy, in its technique and its achievements, has undergone a remarkable transformation; and in the same period practical wireless telephony has sprung into being. These developments have been due to the production and application of the high-vacuum three-electrode thermionic tube or triode*. While wireless telegraphy, under the driving stimulus of the war, has been the home in which the triode was born and nurtured—a home whose customs and practices have been revolutionised by the precocious youngster—it is not only to wireless that its services will be rendered. In every branch of electrotechnics concerned with the detection or measurement of small effects, the triode is destined to become a new instrument of boundless utility, converting the impossible into the possible, and making easy in the field what was difficult in the laboratory. It has been well compared with some fundamental device like the wheel or the lever in mechanics.

A knowledge of the physics of the tube itself is not altogether essential to the experimentalist investigating the various applications of triodes and their circuits; for he may take the tubes available to him and find how they may best be applied without concerning himself with their internal functions. But such a delimitation is obviously undesirable, and in the present chapter, devoted mainly to the thermionic tube itself, some outline of its internal theory is presented as a fitting introduction to the study of its application in wireless telegraphy and telephony.

* A convenient term introduced by W. H. Eccles. "Triode" is much to be preferred to the misleading term "valve," an archaism deriving from the early "Fleming Oscillation Valve," which was, or was thought to be, used as a non-return device.

2. THE ELECTRON THEORY

Modern views as to the nature of electricity and matter have been arrived at mainly by investigations of the conduction of electricity through gases; and the electron theory of electricity to which this study has led lies at the base of an understanding of the thermionic vacuum tube. According to the electron theory, which is something of a return to the early one-fluid theory of Franklin, electricity is a kind of basic substance, a constituent of all matter, and has an atomic structure. The ultimate unit or atom of electricity is called the electron, and an atom of matter contains electrons in number and configuration according to the quality of the matter. The electron is the indivisible unit of electricity—of negative electricity according to the customary convention as to sign—and if an atom of matter contains an excess of electrons it is negatively charged, or if a deficit it is positively charged. Electrons have been marvellously measured and counted and weighed*. A current of electricity consists either of a flow of unattached electrons, or of a flow of atoms or groups of atoms containing an excess or deficit of electrons. Such charged atoms or groups of atoms thus conveying, or capable of conveying, electricity by their motion are called ions.

In conducting material there occurs a constant interchange of electrons between neighbouring atoms, so that at any moment there are numerous relatively unattached electrons flitting back and forth in all directions within the body; and these so-called free electrons will be more or less guided in their paths under the influence of an electric field. A difference of potential maintained between two points in a conductor therefore causes a steady drift of dancing electrons from the point of lower to the point of higher potential. There is no resultant transference of matter through the body, because one electron is precisely similar to another, and the electrons are fed in at the one point of the body at the same rate as they are withdrawn from the other.

Although within any conducting body this interatomic dance of electrons is always proceeding, under ordinary conditions practically none of them escapes across the bounding surface separating the conductor from the surrounding insulator, e.g. the atmosphere or vacuum. When, however, the temperature of the

---

* Mass $= 9 \times 10^{-28}$ gram $= \frac{1}{1850}$ of mass of hydrogen atom.
  Charge $= 1\cdot6 \times 10^{-20}$ E.M. unit.

body is raised, the dance of the tiny electrons, like that of the molecules with which they are in temperature equilibrium, becomes more violent; and with progressive rise of temperature more and more of the electrons at the surface acquire velocities causing them to shoot beyond the control of the atomic attractions and leave the body. There is then an evaporation of electrons, of electricity, from the hot body, somewhat analogous to the evaporation of (say) water at a surface exposed to air. The higher the temperature, the greater the number of electrons emitted per sq. cm per second. The emission depends very much on the surrounding gas, but in a sensibly perfect vacuum it approximately follows the theoretical law of O. W. Richardson

$$N = A\sqrt{T}\,\epsilon^{-\frac{b}{T}}$$

where N = number of electrons emitted per sq. cm per sec.,

T = absolute temperature of the body,

A = a constant varying widely amongst different materials,

b = a constant, roughly 5×10^4 in all materials.

For the metal tungsten, the material of which modern electric lamp filaments are made, commonly used also for the filaments of thermionic tubes, I. Langmuir found

$$A = 1\cdot6 \times 10^{26}$$
$$b = 5\cdot3 \times 10^4.$$

Since 1 ampere is a current of about $6\cdot3 \times 10^{18}$ electrons per second, the formula for tungsten may be written:

Emission current per sq. cm

$$= 2\cdot6 \times 10^7 \sqrt{T}\,\epsilon^{-\frac{5\cdot3\times10^4}{T}} \quad \text{amps. per sq. cm.}$$

This function of T is plotted in Fig. 53 for temperatures in the neighbourhood of those commonly used, e.g. 2,400°. The curve rises very rapidly with the temperature, and for obtaining large emission currents it is very advantageous to use a material, like tungsten, which can be run at a very high temperature*.

* M.P. of platinum = 1,700° C.; of tungsten = 2,800° C.

WATER CONNECTIONS FOR
POLE TIP COOLING COILS

TERMINAL BOARD

EXHAUST
PIPE

WATER COOLED ANODE JACKET

TERMINAL

ANODE CLAMPING RING

ANODE HOLDER

ANODE LOCKING
WHEEL

ANODE DOORS,
MAY BE OPENED
FOR CLEANING
ANODE INSULATING
BLOCK

ARC GROUND
CONNECTION

PLATE XI. FEDERAL-POULSEN ARC, 500 kW (p. 63).
(Federal Telegraph Co.)

PLATE XII. ANTENNA COIL AND SIGNALLING SWITCHES FOR 30 kW ARC (p. 63).
(Federal Telegraph Co.)

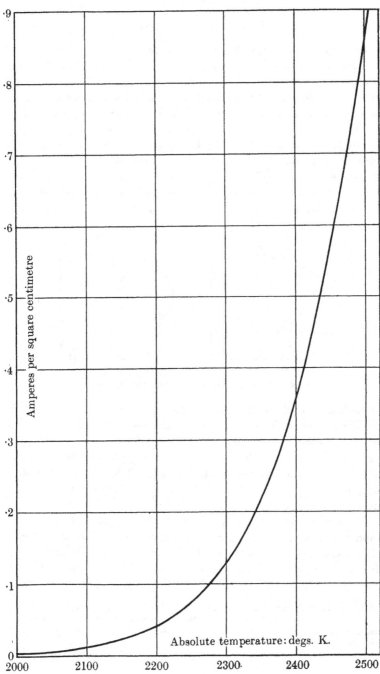

Fig. 53. Rate of evaporation of electricity from hot tungsten.

3. THERMIONIC CURRENTS

In experimenting with, or practically utilising, currents of evaporated electricity, it is convenient to give the hot electrode the form of a filament rendered incandescent by the passage of a current, as in an ordinary electric lamp. This heating current is merely auxiliary and incidental, and must not be confused with the thermionic current. In theory, the electrode might be heated in any way, for example·with a burning glass. The hot filament, or cathode from which the electrons are evaporated, and the other electrode, a cool metal plate or wire, are sealed into a vacuous glass bulb. Although two wires must be brought out from the filament for introducing the heating current, as far as the thermionic current is concerned the whole filament forms a single electrode*.

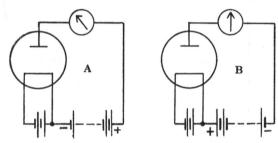

Fig. 54. Unidirectional conduction through vacuum.

In such a tube, if we apply an electric field to the space between the hot and cold electrodes so as to urge the evaporated electrons in the direction hot-to-cold, i.e. if the cold electrode is kept at a higher potential than the hot, a current will traverse the vacuous space, as indicated in Fig. 54 A; but if the field is in the opposite direction, as in Fig. 54 B, no current flows because no electrons are emitted from the cold electrode.

The unilateral conductivity between hot and cold electrodes in a more or less vacuous space was observed by A. Edison in 1883, when experimenting with the then newly invented incandescent electric lamp, and became known as the "Edison effect." It was studied by J. Elster and H. Geitel, and by J. A. Fleming; but

---

* That the filament is not at a uniform potential has, however, sometimes to be taken into account. It is the custom to take the negative end of the filament as the datum line, or zero potential, in specifying the potentials of grid and anode.

became intelligible only in the light of J. J. Thomson's discovery of the electron in 1897 and of the subsequent researches. It remained for I. Langmuir in 1915 to show that the emission of electrons from hot bodies does not require the presence of gas, as had been believed till then, but could occur in a sensibly perfect vacuum. It is to Langmuir's researches with vacua much more perfect than had hitherto been attained that the distinctive quality of the modern thermionic tube is due.

The unidirectional property illustrated in Fig. 54 is responsible for the application of the name "valve" to such a vacuum tube. The tube can be used as an electric non-return device precisely analogous to the leather flap valve of a garden pump or the Dunlop valve of the ordinary bicycle tyre; and with a really good vacuum it behaves as a very perfect valve indeed, barring the way to reversal of current almost completely. This can be illustrated in a striking

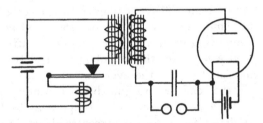

Fig. 55. Experiment with electric valve.

manner by the simple experiment illustrated in Fig. 55. A buzzer or electric bell sends an intermittent current from a feeble dry battery through the primary of a small step-up transformer, to the secondary of which a thermionic tube* and a condenser of 2 or 3 microfarads are connected in series. A spark gap set for 1,000 volts or so is connected across the condenser. The latter gradually becomes charged up—it may take half a minute—until a P.D. is reached sufficient to crash across the spark gap.

Thermionic tubes with a hot cathode (incandescent tungsten filament) and cold anode are used as valves in this way for obtaining D.C. supply at high tension from an A.C. source, and have been constructed for tensions of the order of 100,000 volts, and currents of $\frac{1}{2}$ ampere or more. Such valves are very convenient for providing

* An ordinary receiving triode (Plate XVIII), with grid and anode connected together, may be used.

the H.T. D.C. supply of a 1,000 volts and upwards required in high-power triode transmitters described hereafter.

4. The space-charge

We have seen that the electrons emitted by the cathode would be urged across the vacuum to the anode by an electric field in the direction anode-to-filament. But if the anode is at no higher potential than the cathode, this field is lacking, and the electrons emitted congregate as a cloud very close to the filament. The space around the filament thus accumulates a charge of negative electricity, which thrusts back into the filament the electrons subsequently evaporated by it. This charge is called the "space-charge." It is only if, and so far as, the space-charge is removed that the stream of electrons leaving the filament can continue to exceed the returning stream of electrons thrust back by the space-charge. At any particular filament temperature, a certain number of electrons are emitted every second, but these will be balanced by an equal number of returning electrons unless after emission they are carried away from the neighbourhood of the filament. There is a close analogue in the everyday evaporation of fluids. The puddles in the road dry up quickly on a windy day simply because the water vapour is constantly swept away by the wind; whereas on a still day the vapour congregates above the surface of the liquid and inhibits further evaporation.

In the vacuum tube, therefore, the current between anode and cathode depends on the vigour with which the electrons are swept away from the filament, and therefore increases as the sweeping agent, the electric field between anode and cathode, is strengthened. At every point the strength of the field is proportional to the P.D. between anode and cathode. As the potential of the anode is raised, the current rises and continues to rise until the potential is so great that the electrons are removed as fast as they are evaporated, that is until the current has reached what is called the "saturation" value. After saturation, no further rise of anode potential can augment the current.

Langmuir worked out the theory of the space-charge effect, and showed that, with an unlimited supply of electrons, their rate of flow from cathode to anode is proportional to the $(\frac{3}{2})$th power of the P.D. In the case of an incandescent wire cathode of length

l cm, at the axis of a cylindrical anode of radius r cm, with a P.D. of V volts between anode and cathode, the current is given by

$$I = \frac{14 \cdot 65 \times 10^{-6} \, l}{r} \, V^{\frac{3}{2}} \text{ amperes.}$$

I therefore increases as the $(\frac{3}{2})$th power of V until saturation is reached, and thereafter remains constant. In Fig. 56, this curve is

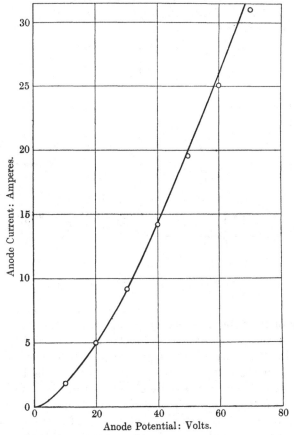

Fig. 56. Foot of diode characteristic. Curve calculated for anode 1·9 cm long, 0·5 cm radius. Points marked observed with "U 3" valve.

plotted with the numerical values corresponding approximately with the anode of one of the tubes we shall be studying later, in which $l = 1·5$ cm and $r = ·5$ cm. To allow for fringing at the ends of the anode, we may add (say) ·2 cm at each end, bringing l to 1·9 cm

The dots shown on the graph are points actually observed in an experiment with a valve of these dimensions*. It is seen that they are in close agreement with the calculated curve.

In accordance with the ($\frac{3}{2}$)th power law, the anode-filament path within the tube is a conductor which does not obey Ohm's law, the characteristic being curved instead of straight, as in a crystal detector (Chapter v, Section 2). It can therefore take the place of a crystal detector (as in Fig. 49, Chap. v) for rectifying wireless signals, and is so indicated in Fig. 57.

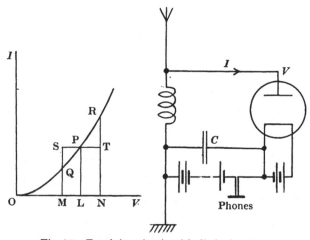

Fig. 57.   Receiving circuit with diode detector.

It is to be noted that we are not here using the non-return property of the tube, but simply the asymmetry of its characteristic about the representative point *P*. Both halves LM, LN of the signal E.M.F. effect a change of current, but the positive change RT is greater than the negative change QS. The tube is therefore not acting as a valve in the sense established in mechanics, and as we have seen it can be used; and the term "valve" is no longer quite appropriate†. If with the weak signals of a wireless receiver the tube were used with such a low anode potential that no current flowed during the negative half-period of signal E.M.F., the rectified current would be vastly smaller.

* The Army "U 3" valve, which is practically the triode of Plate XVIII with the grid omitted.

† "Diode," signifying two-electrode thermionic tube, is preferable. Cf. footnote on p. 78.

J. A. Fleming, in 1904, was the first to use a thermionic tube, a diode, for rectifying wireless signals. His tube was really very much the same thing as that of Fig. 56, the chief difference being that in those days such perfect vacua could not be produced, and the residual gas took some part in the action. The "Fleming Valve" was less sensitive than a good crystal detector, but was used a good deal by the Marconi Company on account of its immunity from damage by atmospherics and other disturbance to which all crystal detectors are susceptible. It is shown at A in Plate XVII. The filament was of carbon or tungsten, and was closely surrounded by a cylindrical anode.

5. INFLUENCE OF GAS ON THE SPACE-CHARGE

Let us return now to the space-charge, the cause of the curvature of characteristic upon which the "detector" action of the diode depends. The thermionic current, i.e. the net rate of emission of electrons from the filament, is determined by the rate at which the space-charge is destroyed; and we have seen that the space-charge may be reduced or even wholly removed by raising the anode potential so as to drag the electrons away to the anode. There are two other agencies by which the effect of the space-charge may be reduced; one is the ionisation of residual gas in the imperfect vacuum; the other is the presence of a positively charged body in or near the space-charge.

If the vacuum is not perfect, collisions must occur between the electrons emitted by the filament and the molecules of the residual gas. Some of these impacts are violent enough to split up the molecule into an electron and a positive ion. The electron will move on towards the anode, perhaps meeting and ionising more molecules on the way; and the positive ion will be driven away from the anode into the space-charge, and perhaps through it into the filament. If the anode potential is high, and therefore the electric field strong, the electrons acquire great velocity, and ionisation of the residual gas becomes visible as a blue glow; but whether the ionisation is visible or not, its effect is to increase the anode current very much. This may be a useful condition, and the most sensitive single tubes have been low-vacuum or "soft" tubes; but the action in such tubes depends so acutely on the gas pressure, which changes by occlusion and the reverse process, and with temperature, that soft tubes required fine adjustment and expert handling. For

example, the triode* produced by H. J. Round of the Marconi
Company, shown at C in Plate XVII, would generally give wonderful results in expert hands; but it had to be carefully nursed, and its
vacuum was actually controlled by the operator from time to time
by warming more or less the pellet of asbestos sealed in a pocket
in the tube, and visible in the photograph. In other tubes, a
mercury amalgam was introduced for a similar purpose.

6. INTRODUCTION OF THIRD ELECTRODE

We pass now to the effect on the space-charge of introducing
into it, or near it, a third electrode, maintained at a higher potential
so as more or less to neutralise the charge. The tube then becomes
a three-electrode tube, and is called a "triode." The third electrode
is usually in the form of a perforated plate, or a wire grid, placed
between cathode and anode, and is commonly called the "grid†."
The introduction of the grid was the work of L. de Forest in 1907,
and although the rationale of the triode does not appear to have
been fully understood at the time, and its uses have only recently
been developed, this addition to the Fleming Valve marks the
conversion of the thermionic tube from a mild convenience into
the potent and indispensable instrument it is to-day.

The reader will be helped to some physical picture of the action
of the grid by studying the diagrams in Fig. 58. Here C, G and A
are sections of the cathode, grid and anode, which may be regarded
for simplicity as portions of large plane surfaces perpendicular to
the paper. The diagrams are intended to show the lines of electric
force in the space between anode and cathode which would act
upon the electrons if any were present, urging them to move in
the direction against the arrow-head (since the electron is a
negative charge). The presence of the electrons modifies the field
in a manner not easy to portray, but by first considering the field
in their absence we can obtain some insight into the action of the
grid. To arrive at the electron currents, which are stated on the
right of the diagrams, we have to picture a cloud of electrons
congregated near the cathode, repelling freshly evaporated electrons

* The fact that this is a triode and not a diode is irrelevant in this connection.
† "Control electrode" (grid) and "repeat electrode" (anode) are good
descriptive terms, used by Eccles. Most commonly, the electrodes are referred
to as filament (or cathode), grid, and anode (or plate).

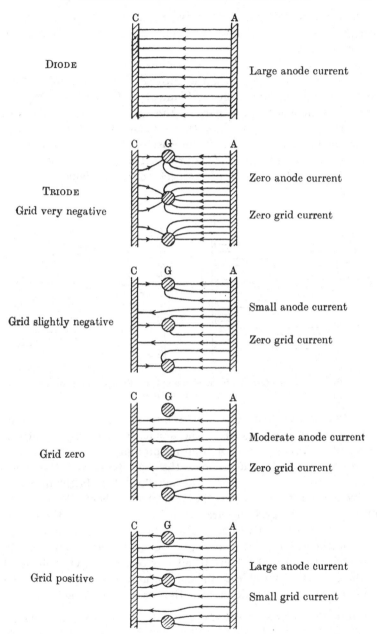

DIODE

Large anode current

TRIODE

Grid very negative

Zero anode current

Zero grid current

Grid slightly negative

Small anode current

Zero grid current

Grid zero

Moderate anode current

Zero grid current

Grid positive

Large anode current

Small grid current

Fig. 58. Electric field between anode A maintained at high potential, and cathode C at zero potential, when grid G is varied in potential. Cathode not heated.

back into the cathode, but ready and waiting themselves to be
dragged away towards grid and anode by an electric field in the
direction A to C.

Below saturation, i.e. as long as more electrons are evaporated
than are carried away from the cathode, no lines of force from
anode or grid terminate on the filament, but each line ends on a
free electron constituting part of the space-charge. In Fig. 59,
an attempt is made to show the lines of force urging the free
electrons towards the anode in the case of the diode, and towards
grid and anode in the case of the triode. The grid is supposed to be
somewhat positive (i.e. with respect to the cathode), and the

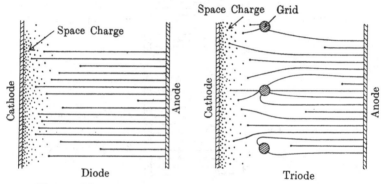

Fig. 59. Lines of force from anode and grid when electrons are
emitted from cathode.

condition unsaturated. It will be seen that the grid to some extent
shields the space-charge from the anode, though only slightly if
the grid mesh is coarse; but from its proximity to the cathode it
exercises a powerful influence over the dense portions of the space-
charge, drawing out electrons and impelling them in the direction
filament-to-grid. Many of these electrons shoot through the
interstices of the grid and are seized by the anode. If the grid is
of very fine mesh, even when positive its effect may be to decrease
the anode current. But in any case the anode current is obviously
influenced by the grid potential, and may be very acutely influenced
if the grid is of suitable form and suitably situated. When the
grid is below the cathode in potential, its current is zero*, and no

---

* If the electrons were released from the cathode with infinitesimal velocity,
this would be strictly true; actually they have appreciable velocities of emission,
so that the grid current does not become zero until the grid potential is a volt
or so negative.

power is absorbed by it; and even when its potential is positive, if the grid wire is fine the grid current is very small. In the practical use of triodes, the grid current is often zero, or so small as to be inappreciable.

7. GRID CONTROL OVER ANODE CIRCUIT

The theory of the internal action of the perfect-vacuum triode cannot be more than sketched in this way here. It has been clearly set forth by W. H. Eccles in the first three numbers of the *Radio Review* (Oct.–Dec. 1919). He there shows theoretically—what Langmuir had already discovered empirically—that in the presence of the grid the unsaturated anode current follows a $(\frac{3}{2})$th power of potential, as in the case of the diode; but that this potential is now no longer the anode potential, but the sum of the anode potential plus a constant times the grid potential. In the diode

$$i_a = A v_a^{\frac{3}{2}}$$

and in the triode $i_a = A' \, (v_a + \nu \, . \, v_g)^{\frac{3}{2}}$

where i_a = anode current,

v_a = P.D. between anode and cathode,

v_g = P.D. between grid and cathode,

and A, A', ν are constants depending on the dimensions and dispositions of the electrodes.

Now the revolutionary possibilities of the triode are wrapped up in this dependence of anode current on the potential function

$$(v_a + \nu \, . \, v_g).$$

This expression implies that the current in the anode circuit is affected as much by a volt of grid potential as by ν volts of anode potential; so that if ν can be made large, changes of current can be effected in the anode circuit by E.M.Fs. introduced into the grid circuit of much smaller magnitude than would be needed if they were applied in the anode circuit directly. This condition alone, however, does not necessarily imply that we have here an amplifier any more remarkable than (say) the ordinary alternating-current step-up transformer (Fig. 60). Here the same change in i_s could be effected by introducing at X a much smaller E.M.F. than would be needed at Y; but we should have to pay, in terms of power, just as high a price at X as at Y. In the triode, on the other hand, we have seen that there is no necessary relation between the effect in the anode circuit and the work done on the grid in producing that effect. The grid current is *not* determined by the anode current; it may indeed be kept sensibly zero, in which case large anode circuit powers may be controlled by changes of grid potential involving the expenditure of no power*.

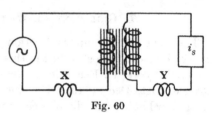

Fig. 60

The characteristic curves between anode current i_a and anode and grid potentials v_a, v_g experimentally determined depart a good deal from the theoretical expression

$$i_a = A' \left(v_a + \nu . v_g\right)^{\frac{3}{2}}$$

owing to various minor considerations omitted from the simplified theory. Amongst these are the edge effects of the electrodes; the necessarily non-uniform potential of a cathode heated by an electric current through it; the not negligible and not uniform velocity of emission of the evaporated electrons; and the non-uniform temperature of the filament, which last is perhaps the most noticeable of all. Nevertheless, the equation

$$i_a = f \left(v_a + \nu . v_g\right)$$

is found to be fairly closely true over wide ranges, although the form of the function f is not the theoretical form stated above. The amplification factor ν therefore remains as the relation between equally effective changes of anode and grid potentials.

8. Typical small high-vacuum triode

The first triode, L. de Forest's "Audion," is shown at B in Plate XVII. The filament was of tungsten, hair-pin shaped. The grid, a coarse wire zig-zag, and the anode, a flat plate, were parallel with the filament. It was a soft tube, and this fact probably accounts for the somewhat barren years intervening between its invention and any adequate development of the theory or application of the

---

* The control over energy transformations exercised by the negatived grid of a high-vacuum triode provides perhaps the most perfect realisable physical analogue of the function of the will in deciding whether a man shall do this or that with the energy within him; neither the grid potential nor, we may suppose, the will entering into the energy equations.

triode. Round's triode, Plate XVII C, was also a soft tube. In it the filament was coated with a mixture of calcium and barium oxides, which possess the property of emitting electrons copiously at a relatively very low temperature*. The grid and anode were concentric cylinders and surrounded the filament. Attention is here confined to the modern form of triode, with sensibly perfect vacuum, the achievement of I. Langmuir and the General Electric Company of America; and in particular to a pattern introduced by the French early in the war, and shown in Plate XVIII.

Here F is the filament, a straight tungsten wire maintained at bright incandescence by a current of about ·7 ampere from a 4 or 6 volt battery; G is the grid, an open helix of nickel wire concentric with the filament; and A is the anode, an outer concentric cylinder of nickel; all being supported on a single glass stem in an exhausted glass bulb. Triodes of this particular pattern are ordinarily used with anode potentials of 30–300 volts and filament emissions of 5–50 milliamps. They were designed to be fairly well suited to a great variety of uses, and with special regard to ease of construction: and their consumption by the French and British during the war must have run into many hundreds of thousands. Great credit is due to General G. Ferrié and the scientists working under his direction at the Télégraphie Militaire for their early recognition of the importance of such triodes and for the early production of this very practical model.

The vacuum in these triodes is very high, usually in the neighbourhood of 1 to 10 millionths of a centimetre of mercury, the final stages of pumping being done with "molecular" pumps of the Gaede (mechanical) or Langmuir (boiling mercury) patterns, and with the aid of liquid air temperature. Moreover great pains are taken to remove traces of gas occluded in, or otherwise held by, the glass envelope and the metal electrodes, which gas might be subsequently liberated to the ruin of the tube. While the tube is on the pumps, the glass is baked almost to softening point; and the anode is made hotter, by electronic bombardment with very high anode potential and overheated filament, than it will ever become in use. By this treatment the residual gas is reduced to such low pressure—and remains so—that collisions between electrons

---

* Oxide coated filaments cannot be employed in high-vacuum tubes because they are not able to withstand the ionic bombardment during the process of evacuation.

emitted by the filament and molecules of residual gas are so few and far between as to have no perceptible influence on the behaviour of the triode. Consequently the old fickle individualities of soft tubes are eliminated; triodes can be manufactured in quantity to a very close specification; and large groups can be used in association without requiring individual electrical adjustments—or, indeed, any adjustments at all.

9. TRIODE CHARACTERISTIC CURVES

A group of typical characteristic curves obtained experimentally with a triode of this pattern is given in Fig. 61. Each of these curves shows how the anode current i_a (or, in one case, the grid current i_g) varies when the anode potential is maintained constant and the grid potential is varied. From a sufficiently complete family of curves of this type, the whole performance of the triode can be calculated. The same information would, of course, be conveyed by a series of curves for each of which the grid potential was kept constant while the anode potential was varied.

The significance of the amplification factor ν can be interpreted by reference to such curves as those of Fig. 61 as follows. Consider, for example, the anode current $i_a = 5$ milliamps. The curves show that this current is produced by the pairs of values of anode and grid potentials shown in the first and second lines of Table III.

Table III.

v_a	500	310	139	86	58	35	10
v_g	-32	$-12\frac{1}{2}$	$+3\frac{1}{2}$	$+9\frac{1}{2}$	$+13\frac{1}{2}$	$+16\frac{1}{2}$	$+26$
$v_a + 10v_g$	180	185	174	181	193	200	270

If the equation
$$i_a = f(v_a + \nu . v_g)$$
holds, a value of ν can be chosen making $(v_a + \nu . v_g)$ the same for each of these pairs of values. The third line of the above table shows that the sum $(v_a + 10v_g)$ is approximately constant, except at very low anode potentials. We therefore deduce for this triode that the equation
$$i_a = f(v_a + \nu . v_g)$$

does approximately apply, and that the amplification factor is $\nu = 10$.

It is to be observed that over a large proportion of their lengths, the curves are approximately straight and parallel. Over this region, the function

$$f(v_a + \nu \cdot v_g)$$

is obviously a linear function, and we may write

$$i_a = A + B(v_a + \nu \cdot v_g).$$

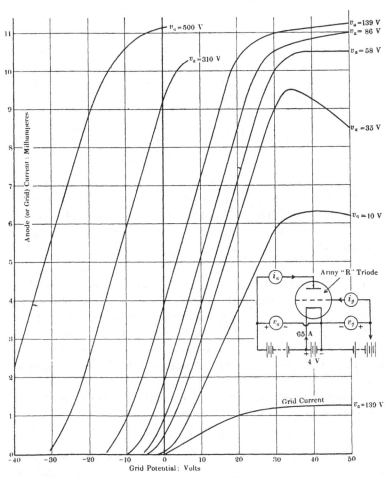

Fig. 61. Typical small high-vacuum triode: observed characteristic curves.

The slope of the curves, viz. the rate of change of anode current with grid potential, which we will write g, is therefore

$$g \equiv \frac{\partial i_a}{\partial v_g} = B \cdot \nu.$$

Similarly the rate of change of anode current with anode potential, which we will write a, is

$$a \equiv \frac{\partial i_a}{\partial v_a} = B.$$

An alternative way of regarding the amplification factor is therefore

$$\nu = \frac{B \cdot \nu}{B} = \frac{g}{a}.$$

Each of these parameters a and g is of the dimensions of a conductance; they may be termed the anode-anode slope-conductance and the anode-grid slope-conductance respectively. They may be separately computed from the curves of Fig. 61, and are there seen to be approximately

$$g = \cdot 33 \, \text{mA/V} = \frac{1}{3,000} \, \text{ohm}$$

$$a = \cdot 033 \, \text{mA/V} = \frac{1}{30,000} \, \text{ohm}.$$

When, as is frequently the case, we are interested only in small changes introduced by a signal or otherwise, we can ignore the steady values of current and potential on which these changes are superposed, and for the regions where the curves are sensibly straight write $\qquad \delta i_a = a \cdot \delta v_a + g \cdot \delta v_g.$

To sum up our deductions from the straight regions of the characteristics of Fig. 61: in considering changes of current in a circuit comprising the path between anode and filament within the tube, this path may be regarded as a conductor whose resistance is 30,000 ohms; and the internal action of the tube is such that the effect of any change of grid potential is the same as that of an E.M.F. of ten times the value introduced into the anode-filament circuit.

A comparison of the characteristic curves actually found with the implications of the theoretical equation

$$i_a = A' \left(v_a + \nu \cdot v_g \right)^{\frac{3}{2}}$$

shows that the concave lower ends agree fairly well. The convex upper ends are due to saturation setting in, and have gradual instead of sharp bends owing to end effects of the electrodes. The

PLATE XIII.   VACUO-THERMO-JUNCTION WITH MILLIVOLTMETER (p. 66).
(Cambridge and Paul Instrument Co., Ltd.)

PLATE XIV. CARBORUNDUM-STEEL DETECTOR (p. 71).
(Marconi's Wireless Telegraph Co., Ltd.)

remarkable straightness of the middle portion seems to be a rather fortunate accident, due chiefly to the fact that the end portions of the filament, contributing a considerable proportion of the total emission, are cooler than the middle portion. Hence saturation sets in at the ends with quite low current and gradually creeps in along the filament as the current is increased.

The curve of Grid Current in Fig. 61 refers strictly to the anode potential 139 volts. In this rather coarse-grid triode, the grid current is somewhat affected by the anode potential; but not very much, and the curve can be taken as showing the grid current approximately with any of the anode potentials. As we should expect from the physics, the grid current is approximately zero unless the grid is at a higher potential than some part of the cathode. It is not quite zero until v_g is slightly negative, as is shown by the enlarged foot of a grid curve in Fig. 62. This is due to the not quite negligible velocities of emission of the electrons. The initial velocity of an electron enables it to reach the grid unless there is an opposing field strong enough to destroy that velocity before it has travelled so far.

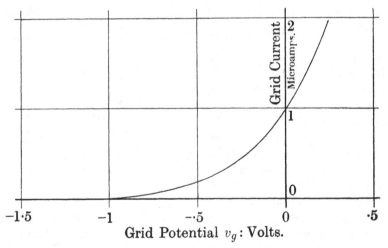

Fig. 62. Foot of grid current characteristic.

CHAPTER VII

THE TRIODE AS AMPLIFIER

1. Elementary single-triode amplification

In the circuit of Fig. 63, a rise of anode current is accompanied by a fall of anode potential equal to the increase of ohmic drop in R; i.e.

$$\delta v_a = - R \cdot \delta i_a$$
$$= - R\left(a\delta v_a + g \cdot \delta v_g\right) \quad \text{[See page 95]}$$
$$\therefore \ \delta v_a (1 + Ra) = - Rg\delta v_g$$
$$\therefore \quad \delta v_a = - \delta v_g \times \frac{Rg}{1 + Ra}.$$

Such an arrangement may be regarded as an amplifier of P.D., in which a rise of 1 unit in the grid potential produces a fall of $\dfrac{Rg}{1 + Ra}$ units in the anode potential. There is no appreciable time lag in this amplifier. Owing to the small mass of the moving parts, the electrons, their velocity is very great: any change of grid potential produces the corresponding magnified change of anode potential almost instantaneously *. Hence this simplest form of triode amplifier, where the only impedances of the circuits are plain resistances, constitutes a truly aperiodic arrangement available for amplifying steady or alternating E.M.F. of any frequency. Its action can be tested very directly with steady E.M.Fs. by observing the change of reading of an electrostatic

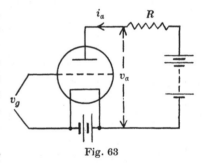

Fig. 63

---

* The velocity acquired by an electron in passing from the filament to the anode at (say) + 70 volts is 5×10^8 cm per sec. Hence if the anode is (say) ·5 cm from the filament, the time taken for the electron to travel across the vacuum is of the order of 2×10^{-9} sec.

voltmeter placed across anode-filament when the grid potential is changed a known amount (say a volt or two), as indicated in Fig. 64.

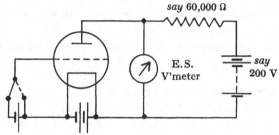

Fig. 64. Measurement of amplification by voltmeter.

2. MAGNITUDE OF AMPLIFICATION OBTAINABLE

The amplification with this arrangement is

$$\frac{Rg}{1 + Ra} = \nu \times \frac{R}{R + 1/a}.$$

This is less than unity (i.e. the output P.D. is less than the input) if $R < \frac{1}{a} \times \frac{1}{(\nu - 1)}$; but it increases asymptotically towards the value $\nu = g/a$ as R is increased to a very large value compared with the anode-anode slope-resistance $1/a$. The maximum amplification obtainable is therefore equal to the amplification factor ν. This, we have seen, is about 10 in the triode of Plate XVIII and Fig. 61. When R is (say) only twice $1/a$, as in Fig. 64, the amplification instead of 10 is

$$10 \times \frac{2/a}{2/a + 1/a} = 6 \cdot 7.$$

The questions immediately arise: Is there any practical obstacle to making the amplification nearly reach the theoretical maximum ν; or to constructing triodes with much greater values of ν than the 10 of this particular pattern? The answer to each of these questions is in the negative, provided only that we have available an anode battery of sufficiently high voltage. The condition for getting its sensibly best performance from the triode is, we have seen,

$$R > > 1/a$$

and with this triode on the steep straight part of the characteristic, $1/a = 30,000$ ohms. Suppose, then, R is made much greater, say

300,000 ohms. With a current of only $\frac{2}{3}$ milliampere, this resistance would itself have a drop of the full 200 volts available; so that the anode current must be less than $\frac{2}{3}$ milliamp. But Fig. 61 shows that the steep straight parts of the curves are not reached until the anode current is upwards of 1·5 or 2 milliamps.; so that below (say) 1·5 milliamps. any increase of R is accompanied by an increase of $\frac{1}{a}$, and the attempt to make $R >> \frac{1}{a}$ tends to defeat itself. If now, the anode battery is raised from 200 to (say) 600 volts, we can obtain an anode current exceeding 1·5 milliamps., and without having a positive grid potential. Thus, with the 300,000 ohms and 600 volt battery, the curves show that the anode current would be about 1·7 milliamps. if the grid potential were about 1 volt negative. R being then as much as 10 times $\frac{1}{a}$, the amplification would be very nearly the maximum ν.

By altering the geometry of the triode, in particular by constructing the grid with a finer mesh, ν can be made as large as desired. For example, the Army "B" triode (used as a low-power transmitting triode) differs very little from the "R," except that the grid has twice as many turns per centimetre; and in this triode, ν is about 18. Again, in the Marconi Co.'s "Q" triode (Plate XVII D), the grid mesh is very fine and the anode diameter large; for it ν is some 60–80. But in these triodes, the minimum anode slope-resistance $\frac{1}{a}$ is much higher, and with finer grid mesh the whole i_a, v_g curve for any particular v_a is shifted bodily to the right; so that an amplification approaching the amplification factor ν could only be obtained by the use of very large R and high voltage.

Now the need for very high voltage in an amplifier is a great practical drawback; hence although theoretically the amplification from a single triode could be made ever so great, such a triode would be very inconvenient to use.

3. Cascade amplifiers

When large amplification is required, the more convenient course is to use two or more triodes in cascade, the anode fluctuations of one triode being impressed on the grid of the next. Then with n triodes, each amplifying m times, the total amplification is m^n. For example, if $m = 8$ and $n = 3$, $m^n = 512$.

In Fig. 65, two triodes each amplifying as in Figs. 63 and 64, are arranged in cascade. The battery b_2 serves to maintain the grid of the second triode at a suitable mean potential, without, of course, affecting the fluctuations of potential to which it is subjected as v_{g1} and therefore v_{a1} fluctuate. The amplification obtainable with this 2-triode combination is the square of the amplification of each

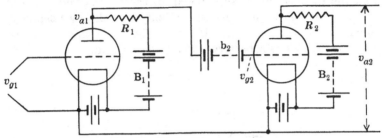

Fig. 65. Conductive connection.

triode separately. With the "R" triode, this might be (say) $8^2 = 64$; e.g. if v_{g1} is made to fluctuate through $\frac{1}{10}$ volt, v_{a2} fluctuates through 6·4 volts; or—a more practical illustration—if v_{g1} varies by 1 millivolt, v_{a2} varies by 64 millivolts.

It is possible to simplify the circuits of Fig. 65 by using a common filament and a common anode battery, as in Fig. 66. An extension

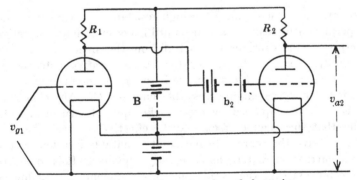

Fig. 66. Common filament and anode batteries.

to three or more triodes, still with common filament and common anode batteries but with separate balancing grid batteries b_2, b_3, ..., is obvious.

The circuits of Figs. 63–66 are suitable, we have seen, for amplifying signals (i.e. changes of potential) of any frequency, including quasi-steady changes. In wireless telegraphy and telephony the signals to be amplified are usually in the form of alternating currents, either of radiation frequency such as tens or hundreds of thousands of periods per second, or of acoustic frequency such as hundreds or thousands of periods per second. For such signals, by the use of transformers and condensers, the steady currents can be separated from the signal currents superposed on them, with very marked practical simplifications. Thus for amplifying ordinary telephonic signals, suitable circuits are shown in Fig. 67, where, it will be seen, the necessity for battery b_2 has been avoided.

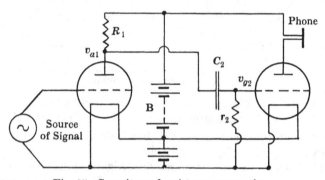

Fig. 67. Capacity-and-resistance connection.

Ignore, in the first place, the high resistance r_2. Grid 2 cannot be permanently positive, because until it became slightly negative electrons would continue to reach it from the filament, removing its positive charge. It is, therefore, at a negative potential of indeterminate value, the difference between its potential and that of anode 1 being taken up across the condenser C_2. Any change in v_{a1} is then accompanied by a precisely equal change in v_{g2}. In order that this change of v_{g2} may be effective, v_{g2} must be of a value to give the representative point a suitable location on the anode current characteristic, as already explained. It is to ensure this condition that the "grid leak" r_2—commonly a megohm or more—is fitted. The grid leak tends to keep the mean grid potential at the desired value, but, owing to the largeness of its resistance, does not much affect the fluctuations of grid potential. There is no upper limit to the permissible value of C_2, since its only function is to protect grid 2 from the steady component of potential of anode 1 while allowing the signal fluctuations to pass. Hence any capacity will do whose reactance at the signal frequency is small compared with the resistance of grid-filament and shunt r_2 (i.e. substantially r_2).

The advantage of this self-adjustment of mean grid potential becomes very great when more than two or three triodes are connected in cascade. With it the voltages of the batteries may vary widely without any ill effect; whereas in a multi-triode amplifier using grid batteries b_2, b_3, ... as in Fig. 66, the effects of very tiny changes of voltage in B, b_2, b_3, ... would be so magnified as to bring the grid potentials of the later triodes of the chain to quite unsuitable values. Multi-triode amplifiers have been constructed (particularly by the French) containing five or more triodes linked by this capacity-and-resistance connection. An advantage of this type of amplifier is its simplicity and compactness, for R, r and C can be easily constructed in forms occupying very little space, and they need not be adjusted to any precise values.

4. TRANSFORMER CONNECTION

In general terms, the function of the resistance R in the anode circuit of the amplifiers so far discussed is, by resisting change of anode current, to make the signal at the grid reproduce itself in the form of the largest obtainable fluctuation of anode potential.

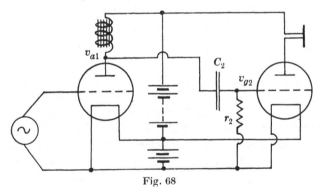

Fig. 68

That the resistance R also resists the steady current is incidental, and unfortunate in that it necessitates an anode battery of higher voltage. For alternating current signals, therefore, R might with advantage be replaced by an inductive choke coil as in Fig. 68. A further advance is made by changing the choke into the primary winding of a transformer, as in Fig. 69. This confers the double advantage of abolishing C_2 and r_2; and, by the use of a step-up transformer, of producing fluctuations of the grid potential v_{g2} actually greater than the fluctuations of v_{a1}.

We are now no longer limited to the theoretical maximum amplification ν, the amplification factor of the triode *per se*. Since the grid current, supplied by the transformer secondary, may be negligible, it might be thought that transformers could be constructed giving a very large step-up of P.D. from primary to secondary; but severe limitations are imposed by the unavoidable self-capacities of the windings of the transformer, and in practice for acoustic frequencies a ratio of turns of only about 1 : 4 is commonly adopted. Even so, mainly owing to this capacity

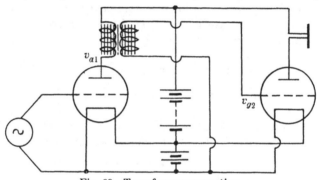

Fig. 69. Transformer connection.

shunting, the amplification per triode does not nearly reach the theoretical limit 4ν.

The design of the inter-triode transformer is of considerable interest, but cannot be examined in any detail here. Suffice it to state that the resistance of the winding hardly comes into consideration, the. fineness of the wire being limited by the difficulty of manufacture and manipulation. The permissible inductances are limited by the self-capacities already referred to. For the very small signal currents, even though they are superposed on large steady currents, the permeability $\dfrac{dB}{dH}$ of the stalloy core is very low—in the neighbourhood of 200. Plate **XIX** shows three patterns of inter-triode transformer which have been extensively used, particularly the small one in the middle. In this, the primary and secondary windings are respectively 4,000 and 16,000 turns of copper wire 2 mils in diameter (S.W.G. 47). For indefinitely small changes of current, such as those corresponding to weak signals, the inductance of the secondary is about 100 henries.

Low-frequency (acoustic) amplifiers of this type were the first important and wide-spread application of the new high-vacuum triodes. They are much used in combinations of 2 to 4 in cascade for amplifying the rectified signals in wireless receivers before they are passed to the telephones. The circuits of such an amplifier are given in Fig. 70. Transformers T_2, T_3 are inter-triode transformers of the type described. T_1 would be suited to the source of signals; and if they are wireless signals from a crystal or other detector D as indicated in the figure, T_1 might well be like T_2 and T_3. T_4 is

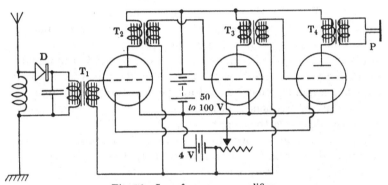

Fig. 70. Low-frequency amplifier.

a step-down transformer protecting the telephones P from the steady anode current*. A common adjustment of filament current is shown, and the anode battery may be provided with tapping points; but neither of these adjustments is really necessary.

It is not customary to connect together more than three or four triodes in this way, mainly owing to the development of amplifier "noises." These are due chiefly to slight unsteadiness within the first triode, apparently depending a good deal on the purity or condition of the filament. These spontaneous fluctuations, after amplification in (say) two subsequent triodes, cause troublesomely loud parasitic rustling and crackling noises in the telephones, and render further amplification useless. The object of these amplifiers is usually not so much to convert an audible into a very loud signal,

* As mentioned on p. 73.

as to render audible an incoming signal which without the amplifier would be quite inaudible. As more and more sensitive amplifiers have become available, they have been used for groping further and further into the hitherto silent field where signals were too weak to be perceived at all. When once these depths have been plumbed so far that the incoming signals are no stronger than spontaneous disturbances originating in the circuits of the first triode, further amplification is of no advantage.

5. HIGH-FREQUENCY AMPLIFIERS

Two types of connection between successive triodes in cascade have been examined, viz. the resistance-capacity circuits and the transformer circuits, both suitable for amplifying signals of acoustic frequency. Provided that the impedances embodied in the circuits —resistances in the one case and reactances of transformer windings in the other—are suitable for the signal currents, the frequency of these currents may be hundreds of thousands instead of hundreds; and the very same circuits which have been described for acoustic frequency amplifiers may be used for wireless frequency amplifiers. The increase of frequency, however, may require profound alteration in the design of the impedances.

It has been pointed out (Chap. I) that the feature of wireless technique differentiating it most strikingly from ordinary (low-frequency) A.C. practice is the importance attaching to very small capacities. A capacity whose reactance is to all intents and purposes infinity for the alternating currents used in electric lighting may offer a very easy path to current alternating at a million periods per second. In constructing large resistance or inductance impedances for currents of wireless frequencies, the difficulty is, therefore, to keep sufficiently small the stray capacities tending to shunt away the high-frequency current. Thus in the resistance amplifier of Fig. 67, if the anode circuit resistance R_1 is 100,000Ω, then at λ = 1,000 metres an accidental capacity across its ends of only about 5 micro-microfarads would carry as large a current as that in the resistance itself. Obviously, for such frequencies, ordinary wire resistance bobbins are unsuitable; they must give place to some form of resistance specially free from capacity.

The anode-circuit resistances (e.g. R_1 in Fig. 67) for high-frequency amplifiers of this type are sometimes made in the form of a thin film of platinum deposited on a plate of glass a few square inches in area, the film being divided into a long zig-zag conductor by finely ruled scratches as in Fig. 71. The grid leaks (e.g. r_2 in Fig. 67), which are subjected to very small P.D. and have to carry very small current, are sometimes constructed as pencil or indian ink lines on paper. A more satisfactory form, due to S. R. Mullard, consists of a partially carbonised cellulose filament enclosed in a protecting tube. A difficulty with all compact high resistances constructed of material (like platinum or carbon) of poor specific resistance is that the cross-section of the conductor

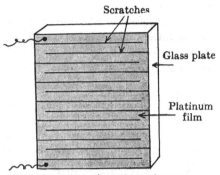

Fig. 71. Platinum film resistance.

must be extremely attenuated, at least in places. The conductor therefore easily loses its continuity, and may develop crazy contacts of more or less microphonic character—a distressing source of amplifier noises.

Not only must the internal capacity of the resistances be kept small, but precautions must also be taken to minimise the objectionable shunting capacities of the wiring, switch fittings, etc., and even within the triode itself. Thus in high-frequency amplifiers of this type for fairly short waves—say below 1,000 metres—special triodes such as the French "horned" pattern (Fig. 72), or the Marconi "V 24" (which is mounted as in the triode shown at D in Plate XVII) are used, to avoid the capacities between anode grid and filament due to the contiguous leading-in wires of the "R" pattern (Plate XVIII).

With these precautions, resistance-capacity connected amplifiers can be made very efficient on long waves; but the efficiency is bound to fall as the wavelength decreases,

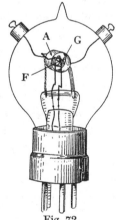

Fig. 72.
"Horned" triode.

since stray capacities cannot be removed altogether and their shunting effect is inversely proportional to the wavelength.

Turning now to the transformer connection of Fig. 69, we find the same difficulty of stray capacities with which to contend. The iron-cored transformers develop into closely coupled induct-ances of usual wireless type—such as single-layer or pile-wound* solenoids, one within the other; or the primary and secondary are wound in inter-leaved sections in a number of close narrow grooves turned in an ebonite cylinder. But again, except with very long waves, the shunting capa-cities prevent the attainment of amplifications per triode nearly as great as the amplification factor v, except near the particular frequency at which the impedance of the inductance and its shunting capacity becomes great as a resonance effect. Thus in Fig. 73, the impedance between X and Y for a steady alternating current of frequency $\frac{p}{2\pi}$ is

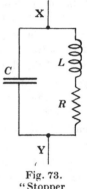

Fig. 73.
"Stopper circuit."

$$\frac{R^2 + p^2L^2}{R + jp\left(CR^2 + p^2CL^2 - L\right)}$$

where $j \equiv \sqrt{-1}$, and so becomes a very great resistance, viz. $\frac{p^2L^2}{R}$,

when $pL = \frac{1}{pC}$ if $R << pL$.

In practice this "stopper circuit" effect is generally utilised. The inductance coils, wiring, etc. are kept as capacity-free as possible; but as much wire is wound on the coils as will make the stopper-circuit effect most pronounced at the wavelength for which the amplifier is most required. The circuits of a triple high-frequency transformer amplifier are given in Fig. 74, for comparison with the low-frequency amplifier of Fig. 70. Con-densers are shown connected across each of the pairs of closely coupled coils, but it is to be understood that these represent only the stray capacities which cannot be avoided. The mean potential of the grids can be maintained somewhat positive by moving the slider of the potentiometer away from the extreme right. This adjustment is provided as a means of controlling the tendency for the whole system to act as a generator of sustained oscillation, and will be understood after the effect of retroaction

---

* "Pile-winding" is an ingenious system of winding on several layers together in such a way as to keep the self-capacity low. The coils visible in Plate XV are pile-wound.

between grid and anode has been studied in Chapters IX and X *.

This type of transformer amplifier can be constructed for quite small wavelengths, although the difficulties of obtaining large amplification except over a very narrow range of wavelengths necessarily increase as the wavelength is reduced.

High-frequency amplifiers have an advantage over low-frequency amplifiers with respect to the spontaneous noises already mentioned (p. 105) as limiting the number of triodes in a low-frequency amplifier to about three. These disturbances, originating within the triodes and perhaps the batteries and resistances, are of a slow

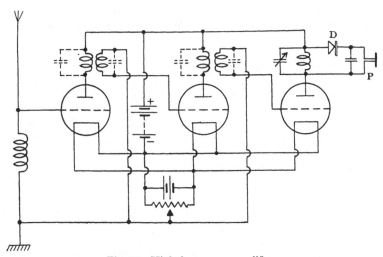

Fig. 74. High-frequency amplifier.

character akin to acoustic fluctuations; they are therefore not conveyed with much amplification through the stages of a high-frequency amplifier in which the connecting condensers or transformers are designed to convey the wireless frequency much more readily than the acoustic frequency. Thus in a high-frequency amplifier with resistance-capacity circuits (Fig. 67), although the connecting condenser C_2 might be of indefinitely large capacity without affecting the sensitivity to wireless signals, if C_2 is kept as small as the wireless frequency will permit, it will not efficiently transmit the parasitic fluctuations of vastly lower frequencies.

* See especially Chapter X, Section 4.

High-frequency amplifiers can therefore profitably be made up with a large number of triodes in cascade; and H. J. Round has stated* that he has used a high-frequency short-wave amplifier containing as many as twenty-two triodes in cascade. Plates **XX** and **XXI** show the exterior and interior of a seven-triode, transformer-connected, high-frequency amplifier suitable for wavelengths between 500 and 1,200 metres. The six high-frequency transformers are clearly seen in Plate **XXI**. The end triode on the left in Plate **XX** acts as rectifier, and is followed by a low-frequency transformer which passes the rectified signal either to a pair of telephones or to a separate low-frequency amplifier.

* Discussion of C. L. Fortescue's paper "The design of multi-stage amplifiers using three-electrode thermionic valves," *Journal Inst. Elect. Engineers*, Vol. 58, Jan. 1920.

CHAPTER VIII

THE TRIODE AS RECTIFIER

1. Anode Rectification

Thermionic currents were first used in wireless telegraphy, by J. A. Fleming, to rectify the received high-frequency signals so as to produce currents perceptible in the telephones. The Fleming Valve, like the crystal detector, depended on the curvature of the characteristic relating current and P.D. This curvature we have already examined in the diode as an effect of the space-charge;

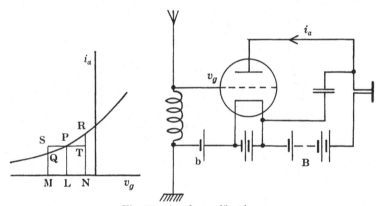

Fig. 75. Anode rectification.

and we have seen that in the triode there is similar curvature at the foot of the characteristic relating anode current and grid potential. Hence if the high-frequency E.M.F. is applied in the grid circuit, and as before the telephones are inserted in the anode circuit, a similar rectifying action will take place. Thus in Fig. 75 (which may be compared with crystal rectification in Figs. 48 and 49 and with diode rectification in Fig. 57), by adjusting grid battery b or anode battery B, the representative point P on the i_a, v_g characteristic can be brought to a place where there is large rate of change of slope. The triode so used gives a signal current in the telephones $\dfrac{e^2}{4} \cdot \dfrac{d^2 i_a}{dv_g{}^2}$. It is a more sensitive detector than the diode

because $\dfrac{d^2 i_a}{dv_g{}^2}$ in the triode is greater than $\dfrac{d^2 i_a}{dv_a{}^2}$ in the diode, a

suitable region of the characteristic being selected in each case*.
A triode, working with a negatived grid joined to the oscillatory
circuit, has the further advantage that no energy is taken from the
oscillatory circuit; whereas the crystal detector and the diode
must take energy from, and therefore increase the damping of, the
oscillatory circuit.

2. CUMULATIVE GRID RECTIFICATION

Another and more effective way of employing the triode as
rectifier makes use of the curvature of the grid current characteristic (see Fig. 62). In Fig. 76, by adjustment of the potentiometer,

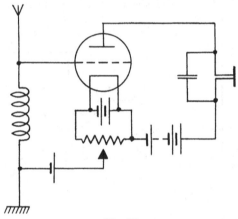

Fig. 76

the mean grid potential can be set to any value shown on the
characteristic in Fig. 62. If it is set to some value lower than about
− 1 volt, no current flows from the grid; but at higher values
of grid potential a current (though a very small one) flows; and
since the characteristic is curved, an alternating E.M.F. from the
antenna produces a rectified current between grid and filament.
This rectified current is very tiny, and in any case does not pass

---

* *Vide* the theory of rectification given in Chapter v, Section 2. The curvature
at the top of the characteristic, due to saturation setting in, may also, of course,
be used to obtain rectification.

PLATE XV. TYPICAL RECEIVER UNIT (p. 77).

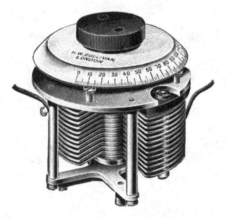

PLATE XVI. CONTINUOUSLY VARIABLE AIR CONDENSER (p. 77).
(H. W. Sullivan.)

through the telephones; so that any rectified current in the telephones must depend, as before (Fig. 75), on the curvature of the anode current characteristic.

But let us now insulate the grid with the condenser C (Fig. 77); and, as in the similar circuit for resistance-capacity connection between amplifying triodes (Fig. 67), provide a high-resistance leak r to prevent the grid from taking up some indeterminate large negative potential. The rectified current is now to some extent integrated to form a charge in condenser C, causing the grid potential to fall more and more as rectification continues under the continued action of the incoming signal. The drop in mean grid potential during a signal is thus cumulative; and indeed, if r were

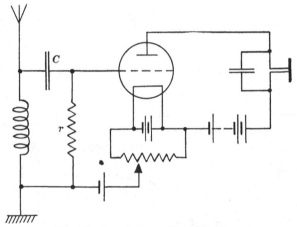

Fig. 77. Cumulative grid rectification.

infinite, after a strong signal the grid might be left at a large negative potential so that the anode and telephone current would be permanently reduced. The function of the leak r may be regarded as that of gradually undoing the work of the signal by letting the electrons accumulated on the grid side of condenser C leak away; so that, after cessation of the incoming signal, the telephone current gradually rises again to its pre-signal value. The proper dimensions for C and r are determined by the conditions that C should be large enough for its reactance to the frequency in use to be much less than the resistance of the grid and its shunt r; but should be small enough to reach appreciably the full P.D. sufficiently rapidly under the action of the signal. The behaviour of a triode rectifying in this way is illustrated in the upper part of Fig. 78.

If the grid is connected through a high resistance (such as 5 megohms) to a point a few volts positive (such as the positive end of the filament), the grid automatically assumes a suitable potential not far from zero, where the current is very small but

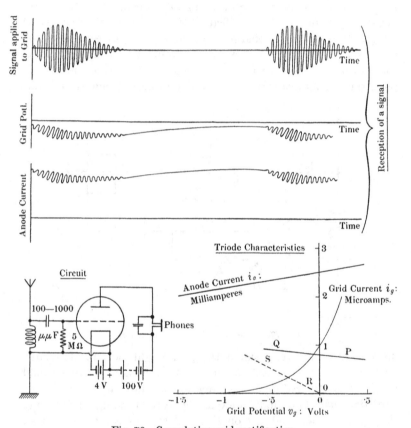

Fig. 78. Cumulative grid rectification.

rapidly changing with rise of potential. A potentiometer adjustment is then unnecessary. In the lower part of Fig. 78, a rectifying circuit of this type is shown, together with the relative portions of the triode characteristics drawn to a large scale. The potential assumed by the grid in the absence of a signal is given by the intersection of the grid characteristic with the line **PQ** whose equation is

$$i_g = \frac{4 - v_g}{r}$$

where r is 5 megohms. If 1 megohm had been used, connected to the negative end of the filament (*vice* 5 megohms to the positive), the grid potential would be given by the intersection of the dotted line **RS** whose equation is

$$i_g = \frac{0 - v_g}{r}$$

where r is 1 megohm. Either of these points would give good rectification of the grid current.

When the signal arrives, the mean potential of the grid decreases, and the corresponding decrease of anode current through the telephones is greater the greater the slope of the anode characteristic. It is desirable therefore to choose an anode potential bringing the representative point on to the steep straight portion of the anode characteristic. Any concavity upwards not only means smaller slope, but also introduces some of the old anode rectification, the effect of which on the telephone current is opposite to that of cumulative rectification.

Thus used, the triode is a much more sensitive detector than either the crystal detector, or the triode detector with simple anode rectification as in Fig. 75; and although it has not the theoretical advantage of the latter of taking *no* current from the oscillatory circuit, the slope-resistance of the grid is so high ($\frac{1}{2}$ megohm or so) that in practice its damping effect on the oscillatory circuit is small*.

* If it were possible to construct oscillatory circuits with indefinitely large ratio $\frac{L}{C}$, the detector conductance would become important however small it might be. The decrement introduced into an oscillatory circuit LC by a condenser leak of high resistance R is $\delta = \frac{\frac{1}{R}}{2nC} = \frac{1}{R} \cdot \pi \sqrt{\frac{L}{C}}$. (Cf. decrement due to small resistance R' *in series* with condenser, $\delta' = \frac{R'}{2nL}$, Chapter III, Section 2.) But the self-capacity of inductance coils imposes practical restrictions on the designer when he attempts to make $\frac{L}{C}$ very great.

3. Relation between rectified current and strength of signal

The quantitative theory of rectification given in Chapter v, Section 2 for the crystal detector applies without modification to the case of triode rectification by curvature of the anode characteristic. The signal current in the anode circuit, due to a high-frequency E.M.F. of amplitude e applied to the grid, is accordingly

$$\frac{e^2}{4} \cdot \frac{d^2 i_a}{dv_g{}^2}.$$

The theory of cumulative grid rectification is not so simple as that of rectification by simple curvature; but it may be shown that the rectified current in the telephone—i.e. the change of anode current—due to a continuously applied signal e is

$$\frac{e^2}{4} \cdot \frac{di_a}{dv_g} \Big/ \left(\frac{1}{r} + \frac{di_g}{dv_g}\right) \cdot \frac{d^2 i_g}{dv_g{}^2}$$

where $\dfrac{di_a}{dv_g}$ is the slope of the anode characteristic, and $\dfrac{di_g}{dv_g}$ and $\dfrac{d^2 i_g}{dv_g{}^2}$ are the slope and rate of change of slope of the grid characteristic (Fig. 78). Hence although cumulative grid rectification may be much more efficient than rectification with crystal or diode, or with triode by dependence on curvature of the anode characteristic, here too the efficiency falls off towards zero as the signal strength is reduced towards zero.

It is this property of detectors which contributes largely to the importance of high-frequency amplifiers. An amplification of (say) 10 times before rectification is worth an amplification of 100 times after rectification; so that although the amplifications per triode practically realised are not so large in high-frequency as in low-frequency amplifiers—particularly with short waves—the use of high-frequency amplifiers allows us to work with incoming wireless signals so weak that without these amplifiers we could not perceive them at all.

4. Heterodyne reception

The heterodyne method of reception has already been described in Chapter v, Section 4. It was proposed by R. A. Fessenden before the triode was invented, but it only came into common use with the advent of the triode as a steady and convenient generator of the local oscillation. The triode as oscillator is the subject of the next chapter; but a further examination of the heterodyne method is appropriate here on account of its bearing on the question of rectifier efficiency discussed in the last Section.

Fig. 79 shows the simplest possible heterodyne arrangement. The rectifier would ordinarily be a triode, but is indicated as a simple crystal detector for simplicity. The form of the local oscillator (in practice invariably a triode arrangement) does not concern us here. In Fig. 80, there is shown:

at (a) the local oscillation of amplitude OA applied to the rectifier;

at (b) the oscillation due to the received signal, of amplitude OB;

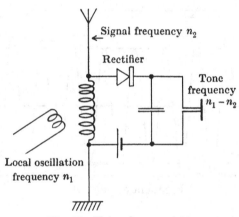

Signal frequency n_2

Rectifier

Tone frequency $n_1 - n_2$

Local oscillation frequency n_1

Fig. 79. Heterodyne receiver.

at (c) the resultant oscillation applied to the rectifier, of amplitude varying at any desired acoustic rate between the maximum $OC_1 = OA + OB$ and the minimum $OC_2 = OA - OB$.
This diagram presents in graphical form the trigonometric summation of p. 74.

The local oscillation serves another purpose besides producing a telephone fluctuation of audible frequency; it also improves the efficiency of the rectifier. That this is so is seen by reference to the rectifier characteristic in Fig. 80, which is shown as consisting of a horizontal portion and a straight inclined portion connected by a curved portion, and which might represent the foot of either the grid or anode characteristic of a triode. We will suppose that the local oscillation is, as it should be, large enough to sweep over the

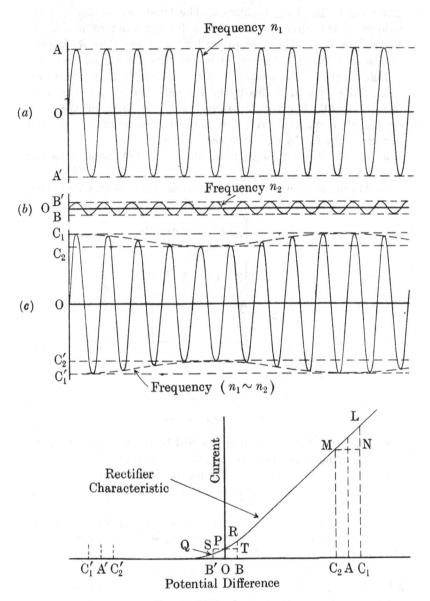

Fig. 80. Heterodyne reception.

rectifier characteristic from the substantially horizontal region on the left to the steep substantially straight region on the right*. The negative peaks of current, being zero, then do not vary at all under the action of the signal, whereas the positive peaks vary (at the acoustic rate) between the values C_1L and C_2M. If there had been no local oscillation, the signal would have produced positive peaks BR and negative peaks B'Q; so that a rough idea of the gain due to the local oscillation is obtained by comparing LN with (RT − QS).

To calculate the approximate magnitude of the "signal current" —here used to connote the extent of the fluctuation of telephone current during the signal—we may proceed as follows. The amplitude of the high-frequency P.D. across the rectifier varies between $(e_a + e_b)$ and $(e_a - e_b)$, where e_b is the amplitude of the

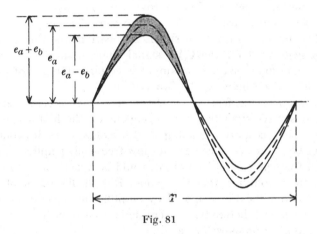

Fig. 81

incoming signal. One cycle of P.D. corresponding to each of these amplitudes is shown in Fig. 81 (as also a cycle of the intermediate amplitude e_a when there is no incoming signal). The ordinates differ most in absolute value, of course, over the middle portion of each half-cycle, i.e. when the grid potential is in the neighbourhoods of A and A' in Fig. 80. Hence if the local oscillation is large enough to sweep a good way over the horizontal and straight inclined regions of the characteristic (Fig. 80), the difference of the quantities of electricity passed under the actions of these two P.Ds. [viz. of amplitudes $(e_a + e_b)$ and $(e_a - e_b)$] is approximately zero during the

* That is, what may be called the uncombined residuum $(a - b)\sin pt$ of the local oscillation (p. 74) is large.

negative half cycle, and during the positive half cycle is approximately the shaded area of Fig. 81 multiplied by the slope m of the straight inclined part of the characteristic. Hence the signal current in the telephones is approximately

$$m \times \frac{1}{T} \int_0^{\frac{T}{2}} \left[(e_a + e_b) \sin \frac{2\pi t}{T} - (e_a - e_b) \sin \frac{2\pi t}{T} \right] dt = \frac{2m}{\pi} e_b.$$

The most important fact brought to light by this expression is that the signal current in the telephones is proportional to the amplitude of the incoming signal, instead of the square of the amplitude as when there is no local oscillation. Hence for very weak incoming signals, the heterodyne vastly increases the sensitivity of the rectifier. It is primarily due to this feature of the heterodyne that such striking improvements in range were noticed in passing from spark to C.W. transmitters. The improvements were due much more to the superior sensitivity of the detector arrangement used for the C.W. signal than in some mysterious transmission superiority of undamped over damped trains of waves manifesting itself between the two stations.

This very useful feature of the heterodyne receiver renders continuous wave signalling less dependent on the high-frequency amplifier than is spark signalling; and if as many triodes could in practice be connected in cascade for low-frequency amplification as for high-frequency amplification, it would be much better to carry out all the amplifying *after* rectifying. But the limitation of low-frequency amplification imposed by amplifier noises still remains; so that even with heterodyne reception it is customary to amplify before rectification as well as after.

5. DETECTING ARRANGEMENTS COMPARED NUMERICALLY

In this chapter three triode detecting arrangements have been examined, viz.

(a) anode rectification (Section 1);

(b) cumulative grid rectification (Section 2);

(c) cumulative grid rectification with heterodyne (Section 4).

With high-vacuum triodes the operation of the rectifier can be analysed in terms of the shapes of the characteristic curves; and for each of these arrangements an expression has been deduced or quoted showing how the signal in the telephone depends on the incoming signal. For comparing the several arrangements, and for pointing the difference between rectification with and without the superposed local oscillation (heterodyne), it is instructive to attach numerical values to the algebraical expressions, by reference to the characteristic curves of the "R" triode. We will write S for the signal in the telephones, S being the excursion of the anode current fluctuating under the action of the incoming signal; and we will write e for the incoming signal, e being the amplitude of the high-frequency P.D. applied between grid and filament. Currents will be expressed in amperes and potentials in volts.

In arrangement (a)
$$S = \frac{e^2}{4} \cdot \frac{d^2 i_a}{dv_g^2}.$$

Examination of the enlarged foot of an anode current characteristic such as those in Fig. 61 shows that the maximum value of $\frac{d^2 i_a}{dv_g^2}$ is about $\cdot 0004$ A/V/V*. Hence $S = e^2 \times 10 \times 10^{-6}$ A about.

In arrangement (b)
$$S = \frac{e^2}{4} \cdot \frac{di_a}{dv_g} \bigg/ \left(\frac{1}{r} + \frac{di_g}{dv_g}\right) \frac{d^2 i_g}{dv_g^2}.$$

Reference to Fig. 78 shows for the case when r is 5 megohms joined to the positive end of the filament that approximately

$$\frac{di_a}{dv_g} = \cdot 33 \times 10^{-3} \,\text{A/V}$$

$$\frac{di_g}{dv_g} = 2 \cdot 5 \times 10^{-6} \,\text{A/V}$$

$$\therefore \; \frac{1}{r} + \frac{di_g}{dv_g} = 2 \cdot 7 \times 10^{-6} \,\text{A/V}$$

$$\frac{d^2 i_g}{dv_g^2} = 4 \times 10^{-6} \,\text{A/V/V}.$$

Hence
$$S = \frac{e^2}{4} \times \frac{\cdot 33 \times 10^{-3}}{2 \cdot 7 \times 10^{-6}} \times 4 \times 10^{-6} \,\text{A}$$

$$= e^2 \times 100 \times 10^{-6} \,\text{A about.}$$

In arrangement (c)
$$S = \frac{2m}{\pi} e.$$

Reference to the grid current curve in Fig. 61 shows that

$$m = \cdot 05 \times 10^{-3} \,\text{A/V}.$$

---

* See the curves in Fig. 6 of the author's paper "The Oscillatory Valve Relay," *Journal Inst. Elect. Engineers,* Vol. 57, April. 1920.

Hence $\quad\quad\quad S = e \times 30 \times 10^{-6}$ A about.

Thus, e volts of incoming signal produces in the telephones

$\quad\quad$ $10e^2$ microamps. with arrangement (a),

$\quad\quad$ $100e^2$ $\quad$,, $\quad\quad$,, $\quad\quad$,, $\quad\quad$ (b),

$\quad\quad$ $30e$ $\quad$,, $\quad\quad$,, $\quad\quad$,, $\quad\quad$ (c).

The values of e in practical use, with and without heterodyne, cover a very great range. In Table IV S is tabulated for $e = 10^{-6}$, 10^{-4}, and 10^{-2} volt by way of illustration.

Table IV.

Incoming signal, volts	Microamps. of signal in telephone		
	Arrangement (a)	Arrangement (b)	Arrangement (c)
10^{-6}	10^{-11}	10^{-10}	3×10^{-5}
10^{-4}	10^{-7}	10^{-6}	3×10^{-3}
10^{-2}	10^{-3}	10^{-2}	3×10^{-1}

Experimental measurements for putting these theoretical results to any precise even comparative test are lacking. Many experimentalists would question the superiority of 10 times shown by (b) over (a); though probably all would agree that the practical figure lies between (say) 1 and 20. It must be remembered that whereas in arrangement (a) the rectifier does not (or need not) introduce any appreciable damping into the oscillatory circuit even with circuits designed to give the greatest practicable step-up in E.M.F. between antenna and rectifier, with such circuits arrangement (b) does produce appreciable damping; so that with the same signal E.M.F. in the antenna, e would be smaller in (b) than in (a).

The superiority of (c) over (b) depends, of course, on the value of e; and the great superiority of the heterodyne method in dealing with very weak incoming signals is clearly illustrated. It is interesting to note that in reporting some direct experiments on this point, L. W. Austin stated*: "Just audible signals were obtained with an input of 12×10^{-6} watt. This is more

* "The measurement of radio-telegraphic signals with the oscillating audion," *Proc. Inst. Radio Engineers*, 1907.

than ... 6×10^5 times that of the normal audion on buzzer signals."

Our formulae
$$S = 100 \ (e_b)^2$$

and
$$S = 30e_c$$

would lead to this power ratio 6×10^5 if
$$e_b{}^2 = 6 \times 10^5 \ e_c{}^2$$

and
$$100 \ (e_b)^2 = 30e_c$$

$$\therefore \ 6 \times 10^5 \ e_c{}^2 = \cdot 3e_c$$

$$\therefore \ e_c = \frac{\cdot 3}{6 \times 10^5} = 5 \times 10^{-6} \, \text{volt}$$

and
$$e_b = \sqrt{6 \times 10^5} \, e_c = 4 \times 10^{-3} \, \text{volt.}$$

It would appear, therefore, that whereas without the heterodyne an incoming signal of 4 millivolts was required to make a just audible sound, 5 microvolts sufficed with the heterodyne.

CHAPTER IX

THE TRIODE AS OSCILLATION GENERATOR

1. SELF-OSCILLATION IN A SIMPLIFIED CASE

An appreciation of how oscillation must arise in the circuits of Fig. 82 when certain conditions are met is a short cut to an understanding of the retroactive principle underlying the triode as oscillator. LRC is a lightly damped oscillatory circuit inserted between grid and filament, and is coupled by a mutual inductance

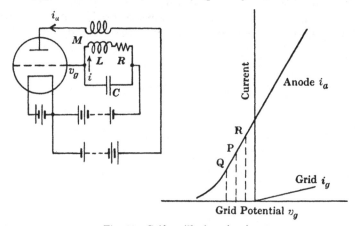

Fig. 82. Self-oscillating circuit.

M to a small inductance inserted between anode and filament. We will suppose the batteries are chosen to place the representative point at P on the anode characteristic, where it is approximately straight and where the grid current is zero.

Let the LRC circuit be disturbed by any means and left oscillating with its natural frequency given by

$$p = 2\pi n \approx \frac{1}{\sqrt{LC}}$$

so that P oscillates between Q and R; and let $\mathscr{I}$ be the R.M.S. and i the instantaneous value of current in LRC. Energy is being removed from the circuit at the rate $R\mathscr{I}^2$, and unless some E.M.F. is impressed from without, compensating for the ohmic drop in R, the amplitude of the oscillation will die down according to the damping factor $\epsilon^{-\frac{R}{2L}t}$. Owing to the reactance drop in the grid coil L, the grid potential v_g fluctuates, lagging $90°$ on the current i and with an R.M.S. value $pL\mathcal{J}$. Assuming that the inductance of, and therefore the E.M.F. in, the anode coil is very small, the anode current i_a approximately follows, in phase, the variations of v_g; so that in the anode coil there is an alternating current $g.pL\mathcal{J}*$ lagging $90°$ on the current in L. This anode coil current induces in the grid coil an E.M.F. $pM.g.pL\mathcal{J}$ in phase or in anti-phase with the current i according as M is positive or negative. If the sense of M is such that a rising anode current induces an E.M.F. tending to raise the grid potential (as is the case with the connections and directions of winding shown in Fig. 82 if the two coils are coaxial), there is introduced into the LRC circuit an E.M.F.

$$p^2gML\mathcal{J}$$

in phase with the current, and therefore contributing energy to the circuit at the rate $p^2gML\mathcal{J}^2$.

The net loss of energy from the LRC circuit is therefore at the rate

$$R\mathcal{J}^2 - p^2gML\mathcal{J}^2$$

or $\qquad\qquad \mathcal{J}^2(R - p^2gML).$

The rate of decay of the oscillation is thus reduced; and if the second term is numerically greater than the first, the rate of decay is negative, and the amplitude of the initial oscillation instead of dying down increases. This means that as soon as such a circuit is brought into being, as by lighting up the filament or coupling the coils together, an oscillation will start and build itself up to some finite amplitude. The final value we do not here investigate.

The limiting condition for an oscillation to start is clearly

$$R - p^2gML = 0$$

i.e. $\qquad\qquad R = \frac{gM}{C}$, since $p^2 = LC$

or $\qquad\qquad M = \frac{RC}{g}.$

---

* $g \equiv \dfrac{\partial i_a}{\partial v_g}$, as in Chapter VI, Section 9, et seq.

Whenever the L.H.S. of this equation is greater than the R.H.S., there is instability and the system sets itself into oscillation.

2. MATHEMATICAL ANALYSIS OF A CIRCUIT

There are many ways of connecting up a triode and an oscillatory circuit so as to produce sustained oscillation in the latter. The principle in all is, of course, to let the oscillatory current so stimulate the grid that a supply of energy is introduced into the oscillatory circuit from the source of power in the anode circuit, viz. the high-tension battery. This is what we have called the retroactive principle. We will study here in some detail one simple arrangement, very commonly used and shown in Fig. 83, in which the oscillatory circuit is in series with the anode instead of the grid as in Fig. 82. The resistance R of the oscillatory circuit covers, of course, not merely the ordinary Ohm's law resistance of the coil, but also all losses of energy from the circuit, such as condenser losses and losses due to coupling to an external circuit where work is required to be done. The actual location or locations of R in the oscillatory circuit is not a matter of any importance; it is shown as

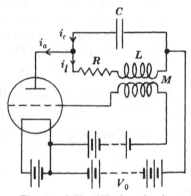

Fig. 83. Self-oscillating circuit.

a part of the inductance coil for simplicity, and because it is usually from the inductance coil that power is drawn to the external circuit.

Let i_c, i_l be the instantaneous values of current in C and L; and let i_a, v_g, v_a be the instantaneous values of anode current and grid and anode potential *changes** from the non-oscillating condition. Then

$$i_c + i_l = i_a = g.v_g + a.v_a.$$

But
$$v_g = M \frac{di_l}{dt}$$

and
$$v_a = -Ri_l - L \frac{di_l}{dt}$$

and
$$i_c = -C \frac{dv_a}{dt} = CR \frac{di_l}{dt} + CL \frac{d^2i_l}{dt^2}.$$

---

* We are concerned with the steady (non-oscillating) values only indirectly, as determining the regions of the characteristics over which the triode is operating.

Making these substitutions in the first equation,

$$CR \frac{di_l}{dt} + CL \frac{d^2i_l}{dt^2} + i_l = gM \frac{di_l}{dt} - aRi_l - aL \frac{di_l}{dt}$$

i.e. $$CL \frac{d^2i_l}{dt^2} + (CR + aL - gM) \frac{di_l}{dt} + (1 + aR) i_l = 0.$$

If $$(CR + aL - gM) < 4CL (1 + aR)$$

the roots of the auxiliary equation are unreal, and the current i_l is oscillatory, viz.

$$i_l = I_l \, \epsilon^{bt} \sin pt$$

where $$b = - \frac{CR + aL - gM}{2CL}$$

and $$p = \sqrt{\frac{1 + aR}{CL} - \frac{(CR + aL - gM)^2}{4C^2L^2}}.$$

According as b is negative, zero, or positive, i.e. as

$$gM \lesseqgtr (CR + aL)$$

the amplitude of the oscillation dies down, remains constant, or increases with time.

The first condition $gM < (CR + aL)$ is used, as we shall see in Chapter x, in retroactive amplifiers, where retroaction between anode and grid circuits is provided in order to decrease the damping of an oscillatory circuit, but not enough to produce self-oscillation. The third condition

$$gM > (CR + aL)$$

is the condition for the oscillation to increase in amplitude, and may be referred to as the "condition for growth." The second or intermediate condition

$$gM = (CR + aL)$$

is the "condition for maintenance" of the oscillation unchanged in amplitude.

The frequency of the oscillation is usually very near to the free frequency of the oscillatory circuit, i.e. approx. $p = \dfrac{1}{\sqrt{CL}}$. The term aR is usually negligible compared with unity; so that actually p is rather less than $\dfrac{1}{\sqrt{CL}}$, i.e. the wavelength of the oscillation produced by the triode is rather greater than the wavelength of the free oscillatory circuit.

3. Applications of such an oscillator

In wireless telegraphy there are two chief fields of utility for an oscillating triode*. It is used as a convenient source of the local oscillation needed for the heterodyne reception of C.W. signals, in which case only very feeble oscillation is required, and the small receiving triodes (e.g. Plate XVIII) are employed with an anode battery of only 50 volts or so. And it is used to produce powerful oscillation in the antenna of a C.W. transmitter for telegraphy or telephony, in which case one or a group of several large triodes is used with a high tension which may amount to thousands of volts. Plate XXII shows a large transmitting triode, known as "T4A," developed by H.M. Signal School (Navy). This tube is used with about 3,000 volts on the anode and can be made to deliver some ¾ kilowatt to the antenna.

In heterodyne oscillators it is not important to obtain large power or high efficiency; but in transmitting oscillators it is; and we proceed now to investigate the state of oscillation arrived at when the conditions necessary for the production of oscillation have been met.

4. Amplitude of oscillation reached

In the foregoing analysis the grid current has been assumed to be negligible. As long as only the inception of oscillation is considered, this is obviously permissible under suitable conditions of working; for by arranging that the initial potential of the grid is negative (as in Fig. 82), the grid current is made actually nil. But as the oscillation grows, there are epochs when the grid is at a positive potential and grid current flows. (A milliammeter in the grid circuit is, indeed, sometimes used to indicate when oscillation starts.) Nevertheless, the grid currents are so much smaller than the anode currents, and still more the power consumed at the grid is so much smaller than the power in the anode circuit, that it is permissible to ignore the effect of the grid current even with large oscillation. This we do in the following examination of what happens in the arrangement of Fig. 83 after oscillation has grown up.

The condition for growth has been found to be

$$gM > (CR + aL).$$

---

* Apart from laboratory high-frequency measurements, in which the oscillating triode is invaluable.

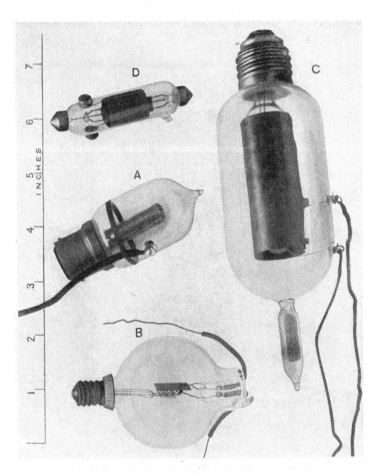

PLATE XVII.

A. FLEMING VALVE (p. 87).   C. MARCONI "N" TRIODE (pp. 88 & 93).
B. DE FOREST "AUDION" (p. 92).   D. MARCONI "Q" TRIODE (p. 100).

PLATE XVIII. FRENCH OR ARMY "R" TRIODE (p. 93).

An oscillation once started therefore increases in amplitude until this condition no longer holds; indeed it builds up until the terms of this expression are so altered in value that the condition of maintenance

$$gM = (CR + aL)$$

is arrived at. The terms which vary as the amplitude increases are the triode conductances g and a; the amplitude reached is therefore determined by them.

Let us rewrite the condition for growth in the form

$$\frac{g}{a}M > L + \frac{1}{a}.CR.$$

Now $\frac{g}{a}$, the amplification factor of the triode, is fairly constant even well up on the bends of the characteristic curves; whereas $\frac{1}{a}$, the anode-anode slope resistance, is a minimum over the straight middle portion of the anode characteristic, but increases towards infinity over the bends at each end. Thus, in Fig. 84, suppose the

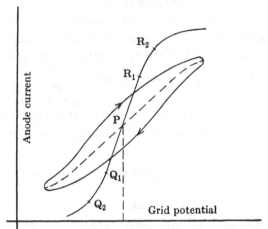

Fig. 84. Oscillating triode (with actual characteristics).

representative point is at P before oscillation starts, and then travels back and forth along the characteristic through that point. It can travel between Q_1 and R_1 without much change in $\frac{1}{a}$; but a wider excursion, between Q_2 and R_2, brings it to regions where $\frac{1}{a}$ (at either or both ends of its travel) is rapidly increasing; and the condition for growth is becoming less fully met*. As the excursion further increases, an amplitude is finally reached where the condition of maintenance

$$\frac{g}{a} M = L + \frac{1}{a} . CR$$

obtains, and the system then sustains steady oscillation at that amplitude.

Actually, however, the representative point does not run up and down a single characteristic curve in this way; for each of these curves is a curve of constant anode potential, and the anode potential fluctuates during the oscillation cycle†. Thus we saw

$$v_g = M \frac{di_l}{dt}$$

$$v_a = - Ri_l - L \frac{di_l}{dt}.$$

So that if we write

$$i_l = I_l \sin pt$$

we have

$$v_g = pMI_l \cos pt$$

$$v_a = - RI_l \sin pt - pLI_l \cos pt.$$

Now pL being very great compared with R, this expression shows that v_a contains a large term pLI_l in anti-phase to v_g, and a small term RI_l leading on v_g by a small angle $\tan^{-1} \frac{R}{pL}$.

If this angle were zero, i.e. if the resistance R of the oscillatory circuit were zero, the representative point P in Fig. 84 would run up and down the dotted line whose straight middle portion is given by

$$i_a = a.v_a + g.v_g$$

$$= - apLI_l \cos pt + gpMI_l \cos pt$$

$$= v_g \left(g - a \frac{L}{M} \right).$$

But since R is not zero, the anode current does not vary quite in phase with the grid potential, and the representative point describes a more or less elliptic closed curve as in the figure. The smaller the resistance of the oscillatory circuit, the more this

---

* When a varies during the cycle, we must regard $\frac{1}{a}$ in the conditions for growth and maintenance as a mean value for the cycle.

† It was to introduce the simplification of constant anode potential that in Section 1 the inductance of the anode coil was taken as very small.

closed loop collapses towards the single line along which P would reciprocate if the resistance were zero.

The amplitude of oscillation is limited by the increase in $\frac{1}{a}$ when the anode current approaches zero or the saturation value; it therefore grows until at least one end of the loop is located in these regions. In order to utilise the greatest possible excursion of grid potential, therefore, the point P should be situated about half way up the characteristic, i.e. the mean grid potential should be chosen to make the non-oscillating anode current about half the saturation current. Since

$$v_g = pMI_l \cos pt$$

this grid excursion will be accompanied by the largest amplitude of oscillatory current I_l if the retroactive mutual inductance M is made as small as possible, i.e. no more than is necessary to sweep

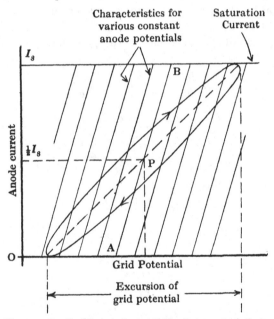

Fig. 85. Oscillating triode (with ideal characteristics).

the anode current from zero to the saturation value. If we make the simplifying approximation that the characteristics are straight lines with sharp instead of rounded corners at top and bottom, the approximate relations of the preceding paragraphs become precise, and we can obtain algebraic results of general application.

Fig. 85 then shows the conditions of working to obtain maximum oscillatory current. Before the retroacting coils are coupled together (i.e. $M = 0$), the grid potential must be adjusted to bring the representative point on the static characteristic AB, corresponding to the anode potential in use, to the mid-point where the anode current is half the saturation current I_s. As M is then gradually increased, immediately it exceeds the limiting value given by

$$\frac{g}{a} M = L + \frac{1}{a} CR$$

oscillation starts, and reaches the full amplitude, causing the anode current to vary between zero and I_s, but not changing its mean value $\frac{1}{2} I_s$. The representative point traverses the ellipse shown. Any further increase of M would increase the excursion of grid potential, causing the ellipse to become a loop with the tips cut off by the zero and saturation current lines; but the value of I_l would diminish.

To find the value of this maximum I_l, we have

$$i_a = g.v_g + a.v_a$$

$$= gM \frac{di_l}{dt} + a \left(- Ri_l - L \frac{di_l}{dt} \right)$$

$$= (gM - aL) \frac{di_l}{dt} - aRi_l$$

$$= pI_l (gM - aL) \cos pt - I_l aR \sin pt$$

$$\therefore I_a \equiv (i_a)_{\max} = I_l \sqrt{p^2 (gM - aL)^2 + a^2 R^2}.$$

Making the substitutions

$$M = \frac{CR + aL}{g}$$

$$p^2 = \frac{1}{CL}$$

this becomes

$$I_a = I_l R \sqrt{\frac{C}{L} + a^2}.$$

But

$$I_a = \frac{1}{2} I_s$$

$$\therefore I_l = \frac{I_s}{2R \sqrt{\frac{C}{L} + a^2}}.$$

This, then, is the greatest obtainable oscillatory current with the given values of L, R and C. It is directly proportional to the filament emission I_s, and is inversely proportional to the resistance R of the oscillatory circuit. The formula indicates also that it would be increased by any decrease of C or increase of L; and since in practice

$$a^2 < < \frac{C}{L}.$$

that it could be so increased to a very great extent. We now proceed to see that this is not always the case.

5. LIMITATION BY INADEQUACY OF ANODE POTENTIAL

The formula for the maximum obtainable oscillatory current I_l was calculated on the assumption that all the anode characteristics in the region of the oscillation (covered by the ellipse) ran up to the saturation current I_s as in Fig. 85. But from our early study of the triode it is obvious that when the anode potential is very low, no extent of rise of grid potential will make the anode current assume the saturation value. The observed curves of Fig. 61 illustrate this fact. If therefore the amplitude of fluctuation of anode potential approaches the P.D. V_0 of the high-tension source, the upper end of the loop in Fig. 85 will be lower than the saturation current. The growth of the oscillation may then be said to be inhibited by lack of anode potential, instead of by lack of anode current as in Fig. 85. Another upper limit to the amplitude is thus given by the equation

$$V_0 = (v_a)_{\text{max.}}$$
$$= pLI_l \text{ approx.}$$
$$= \sqrt{\frac{L}{C}} . I_l.$$

Hence the oscillation ceases to grow either when

$$I_l = \frac{I_s}{2R\sqrt{\frac{C}{L} + a^2}} = \frac{I_s}{2R\sqrt{\frac{C}{L}}} \text{ approx.}$$

or when $\quad I_l = \sqrt{\frac{C}{L}} . V_0$

whichever occurs first. It will continue furthest when $\dfrac{C}{L}$ is chosen to make these two limits coincide, i.e. when

$$\frac{C}{L} = \frac{I_s}{2RV_0}$$

making $\qquad I_l{}^2 = \dfrac{V_0 I_s}{2R}.$

6. PRACTICAL ADJUSTMENTS FOR MAXIMUM OUTPUT

The practical conclusion from the foregoing analysis is briefly this. Given a source of high-tension current of P.D. V_0, and a triode with a certain filament emission I_s, and an oscillatory circuit of resistance R in which we wish to produce as large an alternating current as possible at a specified frequency (or wavelength), proceed as follows:

(i) Uncouple the two coils (i.e. make $M = 0$).

(ii) Adjust the grid potential until the anode current is half the saturation current I_s.

(iii) Gradually couple up the coils until oscillation sets in (and a little more, as a margin of safety).

(iv) Keeping the product CL at the specified value, vary C and L, adjusting as in (iii) on each occasion, until the oscillatory current is a maximum.

According to the theory, the amplitude of this oscillatory current will then be

$$I_l = \sqrt{\frac{V_0 I_s}{2R}}.$$

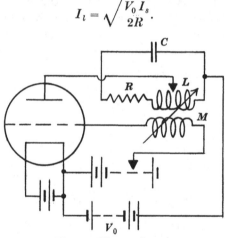

Fig. 86. The anode tap.

It will actually be slightly smaller because of the inaccuracy of the assumption as to shape of characteristic curves; and because the upper limit we have taken for v_a, viz. V_0, can never be reached[*].

In practice it may be very inconvenient to adjust the amplitude of the anode potential fluctuations by varying C and L while

---

[*] With high values of V_0, however, this second departure is not large.

keeping their product constant, as prescribed in (iv) above. Almost the same result is more conveniently obtained by the simple device of providing too small a ratio $\dfrac{C}{L}$ in the oscillatory circuit and including only a variable part of L in the anode circuit, as in Fig. 86. The point at which the anode is joined to the coil is often called the "anode tap." In Fig. 86, moving the anode tap to the left has the same effect as decreasing the ratio $\dfrac{C}{L}$. It consists of one adjustment instead of two, and produces only slight change of wavelength.

7. OUTPUT AND EFFICIENCY

When the system is adjusted for maximum oscillatory current as described, the power consumed in the oscillatory circuit, i.e. the output of the system regarded as high-frequency generator, is

$$R\left(\frac{I_1}{\sqrt{2}}\right)^2 = \frac{V_0 I_s}{4}.$$

The power delivered by the high-tension source, i.e. the input, is

$$V_0 \times \tfrac{1}{2}I_s.$$

The efficiency is therefore 50 per cent.

In obtaining this efficiency, we have treated the whole power in the oscillatory circuit as "output," whereas of course only a proportion of R (though in practice a large proportion) is useful resistance. We have also ignored the auxiliary power consumed in rendering the filament incandescent. The higher the temperature of the filament, and the higher the voltage of the high-tension source, the smaller the ratio between filament power and output power; and in large triodes, under practical conditions, this ratio may be well under 10 per cent. It is to be noted that with the adjustments for maximum output, the mean anode current is the same whether the triode is oscillating or not. The whole of the input is converted to heat within the triode when not oscillating, and half when oscillating.

Since the maximum output of the oscillator is $\dfrac{V_0 I_s}{4}$, the output for any given tube can theoretically be increased without limit by raising the voltage V_0 of the high-tension source. The practical limit is imposed by the heating of the anode. It is customary to load an oscillating triode up to a point where the anode glows a dull red.

For larger outputs, it is then necessary to use a larger triode, or a group of two or more triodes in parallel, i.e. with their anodes, grids and filaments respectively connected together. Provided that the triodes are precisely similar, the group of n behaves like a single triode with the conductances g, a and the saturation current I_s multiplied by n. With the appropriate circuit adjustments, both output and input would be multiplied by n, the efficiency remaining unchanged. Owing, however, to unavoidable dissimilarities between tubes, it is not practicable to keep all the tubes in correct adjustments, and the output and efficiency consequently fall below the theoretical values. It is not usual to group more than six transmitting triodes in parallel.

The construction of the large glass envelopes needed for triodes. capable of dissipating much more power than the 400 watts or so of the "T4A" tube of Plate XXII is very difficult. Hence experimental tubes of higher power have been constructed with envelopes of fused silica, now much used in various chemical manufacturing processes. Silica is, of course, more costly to work than is glass; and the provision of vacuum-tight seals for the leading-in wires is a more serious problem; but cracking under changes of temperature, the great trouble with glass bulbs, is entirely avoided. It seems highly probable that large silica transmitting triodes will soon come into general use.

We have investigated the conditions for obtaining the greatest output from the oscillator, and have found that under these conditions all the fluctuations of current and potential are nearly sinoidal, and that the efficiency cannot exceed 50 per cent. If these conditions are departed from, it is possible to obtain a reduced output with a more than proportionately reduced input, that is a higher efficiency. In particular, for high efficiency it is advantageous to provide a low mean grid potential, bringing the representative point on the static characteristic well down towards the foot of the curve instead of half way up to saturation. The anode current is then no longer sinoidal, but remains zero for nearly half the cycle. A grid battery, as included in Fig. 86, may be used for this purpose; the anode current is then very small until oscillation starts. An alternative arrangement, in which the grid potential is lowered automatically without the use of a grid battery, and which is largely self-compensating for any changes in the high-tension voltage or filament temperature, is shown in

Fig. 87. It is the same arrangement as is used for cumulative grid rectification (Fig. 78). R_g is a large resistance—such as 30,000 ohms—shunted by a condenser as a by-path for the high-frequency

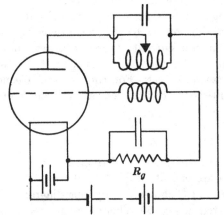

Fig. 87. Oscillating circuit with grid resistance.

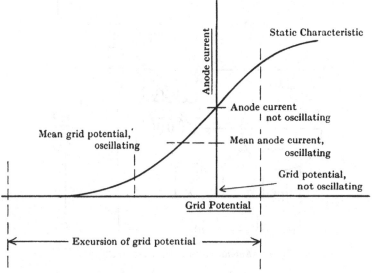

Fig. 88. Oscillation with grid resistance.

component of the grid current. Owing to the rectifying action of the grid, the mean grid current flowing through R_g establishes a negative mean potential, greater in numerical value as the oscillation is more powerful. With a suitable value of R_g, the grid auto-matically assumes the desired negative potential as soon as the oscillation starts (e.g. when the high-tension supply to the anode is switched on). This is illustrated in Fig. 88.

Efficiencies as high as 75 per cent. can be obtained when working in this manner; but owing to the large departure from sinoidal fluctuation of anode current, there is a tendency to produce powerful harmonics in the oscillatory circuit, particularly when this circuit is an antenna*. For this reason, a condition giving an efficiency intermediate between 50 per cent. and 75 per cent. is commonly chosen.

8. THE TRIODE TELEGRAPH TRANSMITTER

When a triode oscillator is used to excite an antenna, the latter may be a separate circuit coupled to the oscillatory circuit with concentrated L and C shown in previous diagrams. This arrange-ment is illustrated in Fig. 89. There are then two separate oscilla-

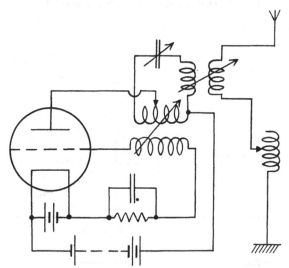

Fig. 89. Antenna with independent oscillator.

tory circuits to be adjusted for wavelength, two mutual inductances, and the anode tap; and power is wasted in heating two oscillatory circuits instead of one. The simplification is therefore usually made

* See next Section.

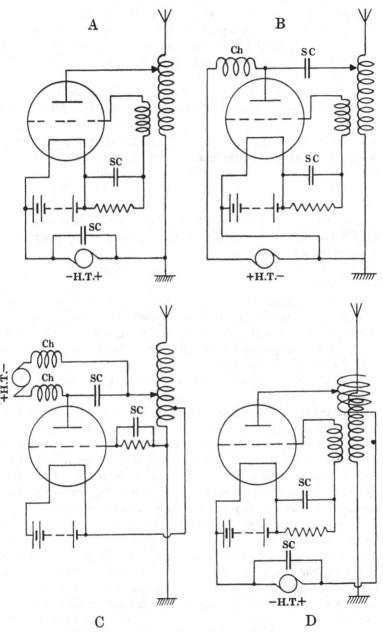

Fig. 90. Telegraph transmitter circuits.

of replacing the closed oscillatory circuit by the antenna itself, as in Fig. 90 A. SC is a high-frequency short-circuit, or shunting condenser to carry the high-frequency component of the anode or grid current, and should be of sufficiently large capacity to offer no appreciable impedance to these currents.

A practical inconvenience of the arrangement shown in Fig. 90 A is that the potential of the filament battery differs from earth by the high tension V_0. Alternatively, if a condenser is inserted in the earth lead and the filament battery is earthed, the mean potential of the whole antenna is V_0 above earth, which is also objectionable. This is avoided in the modification shown in Fig. 90 B, where Ch is a choke coil to prevent the high-frequency component of anode current from being diverted from the antenna coil to which it passes through the condenser SC. The reactance of Ch should be so great that no appreciable high-frequency current traverses it.

Fig. 90 C is a variation in which the separate grid coil is dispensed with.

In unusual cases where it is inconvenient to have as large an inductance inserted in the antenna as would be needed to give the proper anode tap in the circuits of Fig. 90 A, B and C, the anode coil may be a separate larger inductance closely coupled to the antenna coil, as shown at D.

In very low-power transmitters where only two or three hundred volts is required, the source of high-tension supply is usually a battery of dry cells or accumulators, or a D.C. lighting circuit. For larger powers a special high-tension dynamo is sometimes used. But above a few thousand volts, the commutation and insulation difficulties in a dynamo become very great. An A.C. supply with rectifying valves is then used, as in Fig. 91. In this figure, two valves are shown, allowing both half-cycles of the A.C. supply to be utilised. T_1 is a high-tension transformer fed from the alternator A and delivering current to the two anodes alternately. Unidirectional pulses of current, at a frequency twice that of the A.C. supply, pass into the condenser C_1, and thence through the smoothing-out choke Ch to the second condenser C_2 and the external circuit. The filaments are conveniently heated from the same A.C. source through the small auxiliary transformer T_2. Plate XXIII shows a photograph of a large valve developed in H.M. Signal School for this use. The filament emission is over half an ampere; for currents less than this in the direction anode-to-filament the valve offers a back E.M.F. of something under 300 volts, while opposing a sensibly infinite resistance to the passage of current in the reverse direction.

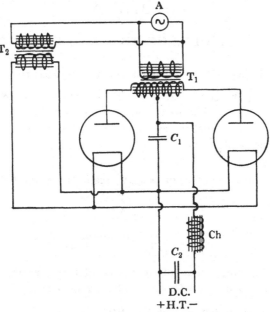

Fig. 91. High-tension D.C. from A.C. source.

The keying of a high-power transmitter is a very serious problem when a generator of the spark, alternator or arc type is used; but keying presents no serious difficulties with triode oscillators. There are obviously many ways in which the oscillation can be stopped and started by means of a Morse key. With small powers, simple interruption of the high-tension supply is often adopted; but with high powers it is preferable to place the key in the grid circuit. For example, in any of the circuits shown in Fig. 90 we may sever the connection between the grid and its inductance coil, in which circuit only a relatively small power is being conveyed. Or we may sever the high-resistance path connecting grid to filament, which is traversed only by a small direct current; the grid then falls in potential by intensification of the rectifying action, usually far enough to stop the oscillation.

CHAPTER X

RETROACTIVE AMPLIFICATION AND RECTIFICATION

1. REDUCTION OF DAMPING BY RETROACTION

IN examining the conditions for a retroactive circuit to generate oscillation, we saw that in the arrangement of Fig. 83, any disturbance in the oscillatory circuit LRC continues as a decreasing or increasing oscillation

$$i = I\epsilon^{bt} \sin pt$$

according as b is negative or positive respectively; and that

$$b = -\frac{CR + aL - gM}{2CL}$$

$$= -\left(\frac{R}{2L} - \frac{gM - aL}{2CL}\right).$$

Thus the damping exponent of the oscillatory circuit, owing to the latter's association with the triode, is reduced from its inherent value $\frac{R}{2L}$ by an amount $\quad \frac{gM - aL}{2CL}$

and can be made as nearly zero as is desired, for example by increasing M. The same fact is expressed by saying that the resistance of the oscillatory circuit is reduced from its inherent value R by an amount $\quad \frac{gM - aL}{C}$.

Any other retroactive circuit capable of generating oscillation, e.g. that of Fig. 82, can be used in the same way as a reducer of resistance when the retroaction is not pushed far enough to generate.

We have, then, in the triode a means of reducing the resistance of any oscillatory circuit to any extent; and since the current built up in a tuned oscillatory circuit by a continued isochronous alternating E.M.F. is inversely proportional to its resistance, it is theoretically possible to produce finite current changes under the action of infinitesimal impressed E.M.Fs. Thus in Fig. 92, LRC is

an oscillatory circuit in which a small E.M.F.—say 1 microvolt—
is impressed from a sustained C.W. signal of frequency n in the
antenna. If the dimensions of the circuit are $L = 5,000$ micro-
henries, $R = 100$ ohms, $C = 508$ micromicrofarads, and therefore

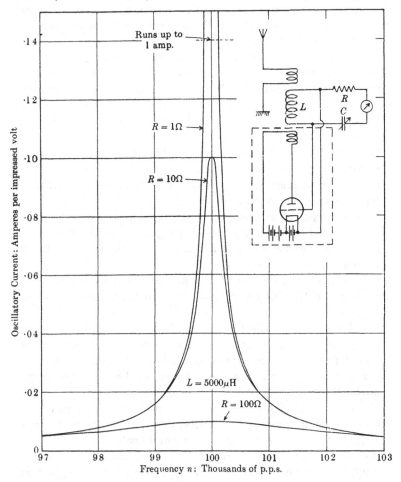

Fig. 92. Resonance curves.

the natural frequency is about 10^5 periods per second ($\lambda = 3,000$
metres), as n is varied, the current varies between a small value
and a maximum of ·01 microampere as indicated in the resonance
curve marked "$R = 100\Omega$"*. If now the triode circuits within

* See Chapter III, Section 2.

the dotted rectangle are connected to the oscillatory circuit, the net resistance of the latter can be reduced to some value much smaller than its inherent 100 ohms, say to 10 ohms or even to 1 ohm. The resulting enhancement of the sharpness of tuning, and the increase in value of the tuned current produced by the same impressed E.M.F., are shown by the resonance curves marked "$R = 10\Omega$" and "$R = 1\Omega$."

Since we are able actually to pass through the zero value of effective resistance of the oscillatory circuit, it might seem at first sight that the engineer has here within experimental grasp the infinities of mathematical conception. But although the device is in practice of extraordinary potence, it fails to achieve infinite sensitivity for the following reasons:

(a) Slight inherent unsteadiness of triodes, batteries, etc. demands the provision of a margin of stability between the actual effective resistance and the region of self-sustained oscillation lying beyond zero resistance.

(b) The negative component of the total resistance—the part introduced by the retroaction—is in general itself affected by changes in potentials of grid and anode, and these vary with the current in the oscillatory circuit.

(c) Extreme constancy of frequency of the incoming signal is necessary if the full benefit from very low resistance is to be obtained.

(d) Although with theoretically perfect conditions the ratio between current and impressed E.M.F. becomes indefinitely great, an indefinitely long time is required before this result is achieved; and in telegraphy the signal is not continued beyond the length of the Morse dot*.

As a means of obtaining oscillatory circuits of extremely low decrement for purposes of selectivity, retroaction may be particularly valuable because no amount of amplification without retroaction can produce the same result; and it has the further practical advantage of giving, with only one or a few triodes, as great an amplification as would be given by many triodes in cascade without retroaction. In order to obtain very large amplification without resorting to excessively small decrement, retroactive amplification

* At 125 words per min. the Morse dot lasts only $\frac{1}{100}$ second. Much more rapid fluctuations still are necessary in wireless telephony, so that here it is of no avail to push retroaction very far.

PLATE XIX. ACOUSTIC-FREQUENCY INTER-TRIODE TRANSFORMERS (p. 104).

PLATE XX. H.F. AMPLIFIER, 500–1200 METRES (p. 110).
(Marconi's Wireless Telegraph Co., Ltd.)

PLATE XXI. INTERIOR OF H.F. AMPLIFIER, PLATE XX (p. 110).
(Marconi's Wireless Telegraph Co., Ltd.)

may be combined with non-retroactive amplification. Thus the triode shown in Fig. 92 may be used as there indicated to reduce the decrement of the oscillatory circuit, and a separate non-retro-active single or multi-stage amplifier may be added to amplify the changes of P.D. occurring across L and C. Alternatively, retro-action may be provided across a whole chain of individually non-retroactive triodes in cascade, as in Fig. 93. Here the antenna

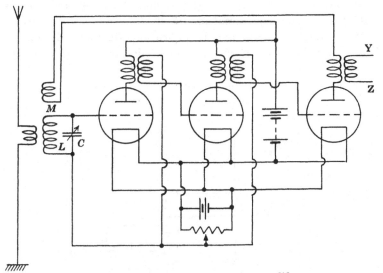

Fig. 93. High-frequency retroactive amplifier.

may be supposed to be loosely coupled to the oscillatory circuit LC, so that a signal received by the antenna impresses a tiny E.M.F. in the circuit LC. The effective resistance of this circuit being reduced as much as is practically convenient by adjustment of the mutual inductance M, the oscillation applied between grid and filament of the first triode is both larger and more sharply tunable as the result of the retroaction. After ordinary cascade amplification in the chain of triodes, the much amplified copy of the original E.M.F. impressed in LC is available at the terminals YZ. This type of retroaction is provided in the high-frequency amplifier of Plates XX and XXI. The sliding handle seen in these Plates shows how the inductive retroaction is there controlled.

2. THE RETROACTIVE RECTIFIER

The ammeter showing the current in the *LRC* circuit of Fig. 92 is, of course, not really used as a detector or measurer of the signal. If no subsequent high-frequency amplification is to be introduced, the detection would be accomplished with a rectifier and telephone joined across *L* or *C* in the ordinary way. But the arrangement of this figure lends itself to the use of the same triode for reducing the resistance of the oscillatory circuit and for detecting the oscillatory current produced. In a triode, the functions of rectification (if any) and amplification are mingled; and in Fig. 94 are shown the slight changes needed to convert the mere retroactor of Fig. 92 into combined retroactor and rectifier of the type shown in Fig. 78 (cumulative grid rectification).

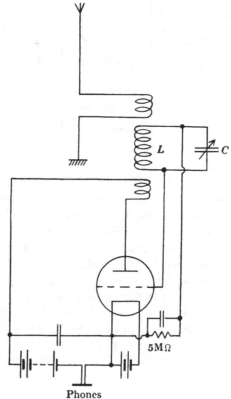

Fig. 94. Retroactive amplifier-rectifier.

3. THE AUTOHETERODYNE

For heterodyne or beat reception, we have seen, all that is necessary is somehow to add to the rectifier a steady oscillation of ɪrequency differing from that of the incoming signal by a suitable small amount. In order to avoid the complication of an independent heterodyne oscillator, we may generate the heterodyning oscillation within the circuits of Fig. 94 themselves. The retroaction is pushed far enough to produce a small sustained oscillation in LC; and to get the beat note the circuit is slightly distuned from the incoming signal. The arrangement is then termed an "autoheterodyne."

With long waves, the distuning may be enough to lessen very seriously the sensitivity of the receiver; but with short waves there is no sensible loss. Thus if the wavelength is 1,000 metres, corresponding with a frequency of 300,000 p.p.s., in order to get a beat note of 1,000 p.p.s. the receiver must be distuned by 1,000 in 300,000, i.e. by only ·033 per cent. The autoheterodyne has a weighty advantage, in some circumstances, over the independent heterodyne. In the former, only one tuning process is involved; in the latter, both of two separate circuits must be fairly precisely tuned before any signals are heard.

4. CAPACITY RETROACTION

It is obvious that any connection between anode and grid circuits which allows the anode oscillation to stimulate the grid with suitable phase relation is competent to introduce negative resistance, and, if pushed far enough, to produce self-oscillation. We have hitherto examined only inductive retroaction; but oscillation may be produced by a capacity connection between anode and grid—a phenomenon, indeed, which is very troublesome in multi-triode high-frequency amplifiers. In Fig. 95, let there be no

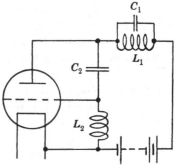

Fig. 95. Capacity retroaction.

mutual inductance between the anode and grid coils L_1, L_2; but let a capacity C_2 be connected across anode-grid. The anode potential fluctuations v_a produce current i_2 through C_2L_2; and i_2 leads or lags by 90° on v_a according as

$$\frac{1}{pC_2} \gtrless pL_2$$

i.e. according as C_2L_2 is below or above resonance. In the former alternative, the grid potential fluctuates in anti-phase to the anode potential, and to a greater and greater extent as C_2 is increased towards the resonant value. We have seen (Chap. IX, Sections 2 and 4) that this anti-phase relationship between v_a and v_g is conducive to self-oscillation.

This type of retroaction is sometimes used intentionally for generating feeble oscillation, as in heterodyne oscillators. Its chief importance, however, lies in the fact that it cannot be altogether avoided, since the capacity between anode and grid within the tube, and between their external connections, constitutes an unavoidable part of the capacity shown as C_2 in Fig. 95. The tendency for oscillation to be set up by this cause increases with the number of triodes connected in cascade—particularly with the transformer connection—owing to capacity coupling between the early triodes and those towards the end of the chain where the fluctuations are relatively very violent. It is for this reason that in Fig. 74 provision is made for adjusting the potentials of the grids to some positive value. The whole system is damped in this way just enough to inhibit self-oscillation. In effect, adjusting the potentiometer in Fig. 74 so as to raise the potential of the grids, is equivalent to lessening or reversing the retroaction coupling due to M in Fig. 93. In the high-frequency amplifier of Plates XX and XXI, the grid potentiometer is seen on the right in Plate XX. A controllable inductance retroaction is also provided in this instrument, varied by means of the sliding handle.

The mathematical theory of multi-triode high-frequency amplifiers is not simple. It has been well stated by C. L. Fortescue in a paper entitled "The design of multi-stage amplifiers using three-electrode thermionic valves*."

5. RESISTANCE RETROACTION

It is obvious that with a chain of two or more triodes, circuits can be devised in which resistances take the place of reactances

* *Journal Inst. Elect. Engrs.* vol. LVIII, Jan. 1920.

in providing the retroaction necessary to approach the self-oscillating condition. Internal resistance in the common anode battery provides such retroaction, and is sometimes responsible for the "howling" of low-frequency amplifiers. Thus in the resistance amplifier of Fig. 96, if the anode battery has resistance R,

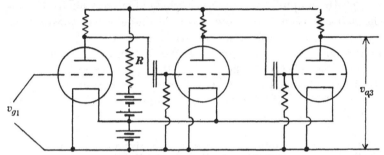

Fig. 96. Resistance in common anode battery.

a change of anode current δi_3 in the third triode produces a change of potential $- R\delta i_3$ at the anode of the first triode and therefore at the grid of the second; and this augments further the change in i_3, and so makes for instability. A large condenser bridged across the anode battery and its resistance would reduce this retroactive effect.

Resistance retroaction has not yet found much application; but

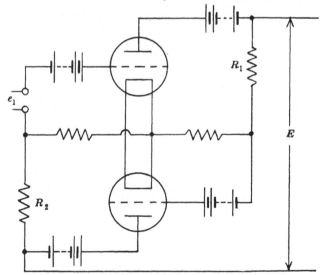

Fig. 97. An amplifier with resistance retroaction.

it possesses considerable interest as giving truly aperiodic effects, which cannot be obtained when transformers and condensers are used. Thus the two-triode amplifying circuit shown in Fig. 97 * was found capable of giving amplification $\dfrac{E}{e_1}$ of as much as 2,000 (although the amplification in either triode alone was about 3); and the signal e_1 can be of any form, including such quasi-steady changes as a long dash on a submarine cable. The action is easily understood. A rise e_1 of grid potential in the upper triode causes an increase of current flowing in R_1, and therefore a fall of potential at the lower grid. This reduces the current in R_2, and so tends to raise the upper grid potential. Any incoming signal e_1 thus reinforces itself.

* See "The Kallirotron, an aperiodic negative-resistance triode combination," by L. B. Turner. *Raaio Review*, vol. I (1920).

CHAPTER XI

WIRELESS TELEPHONY

1. Telephony and Telegraphy Compared

WHILE the broad outlines are the same in both, the smaller details of theory and design take a more prominent place in telephony than in telegraphy. The way of the experimentalist is harder, and the need for theoretical investigations is more acute. Indeed, telephony involves every point of theory and every practical difficulty (except the use of the Morse code) met with in telegraphy, with many others added. It has been well observed that "the difference in degree is not far from that between ruling a dot-and-dash line and making a dry-point etching of an autumn landscape*." And while the methods of wireless telegraphy, in their broad aspects, have reached some stability of form, the art of wireless telephony is still so immature, and new proposals pour out in such a flood, that a statement of the common practice of to-day is hard to make, and a prediction of the methods and achievements of a few years ahead impossible.

But there is no room to doubt that wireless telephony has now emerged from the stagnant position it occupied for years before the advent of the triode oscillator, when freak or demonstration ranges of 100 miles or so were obtained at intervals without offering any prospect of commercial adaptation. The sensational demonstration now takes the form of transoceanic telephony—sensational indeed, since there was no prospect of thousand-mile submarine cable telephony ever becoming feasible. Wireless telegraphy over hundreds of miles now presents no difficulties, and inter-aeroplane telephony had reached the practical stage before the armistice in

---

* A. N. Goldsmith, *Radio Telephony* (1918). This book contains a good descriptive account, very well illustrated, of the development of wireless telephony from the beginning. There is no attempt at mathematical analysis, but it should be a useful introduction to a deeper study of the subject. See also P. R. Coursey's *Telephony Without Wires* (1919), in which a lengthy bibliography is included.

1918. Plate XXIV shows the general appearance a complete wireless telephone may assume.

That wireless telephony is destined to become an important means of communication in the near future is certain; and that the wireless system will be linked with the land-line system is probable. Although striking technical developments are to be confidently expected, even now the provision of commercial services must be regarded as waiting only upon demand and organisation. A sketch of the methods of wireless telephony is therefore included in this book; but it must be understood that only a somewhat arbitrary selection is made, with the object, in this chapter as elsewhere, of introducing the reader to the principles involved. The subject calls urgently for theoretical and experimental investigations, and its literature is mainly in the form of patent specifications in which the chaff and the wheat have yet to be separated.

As a subdivision of wireless telegraphy in the larger (and legal) sense, wireless telephony is distinguished by the following features:

(a) The transmitter must be of the continuous wave type; or if a true continuous wave is not produced, the period of fluctuation of amplitude (or the interval between consecutive separate wave trains) must be so small as to correspond with a "tone" which is preferably of inaudibly high pitch, or which is at least of higher pitch than the highest important constituent frequency of articulate speech.

(b) The amplitude or the frequency of the oscillation in the transmitting antenna must be controlled by some form of voice-sensitive microphone, instead of a Morse key.

(c) The receiver must contain a telephone sounder* as the final indicator of the signal.

(d) Beat heterodyne reception must not be used†.

The only one of these four features of telephone apparatus which is not familiar to the student of telegraphy as already outlined in this book is (b). The same C.W. generators—arc, alternator or

---

* The term "telephone sounder" is here used to connote the ordinary Bell's instrument one applies to the ears. In Germany and France this is called a telephone, and in England sometimes a telephone receiver; but both these terms are also commonly used in their wider senses, and must be so used in this chapter.

† But for improving rectifier efficiency (see Chapter VIII, Section 4) a local oscillation, tuned dead-on, may be provided.

triode—as are used for telegraphy are used for telephony. The telegraph receiver practically always includes a rectifier and telephone sounder; and although such sounders are designed rather to be sensitive at about 1,000 periods per second than to give good voice articulation, they are always usable for telephony. Hence any wireless telegraph receiver as used for spark signals—i.e. without a heterodyne—may be used as a wireless telephone receiver, the rectified current in the sounder increasing and decreasing with the microphonic modulation at the transmitter. In this chapter we are concerned, therefore, mainly with the means by which the transmitting voice is enabled to control the antenna oscillation.

2. Microphonic control at the transmitter

The earliest and simplest method of applying a microphone to modulate the amplitude of the transmitted waves was to insert it in the antenna itself, or in a circuit coupled to the antenna, as in Fig. 98, where M is the microphone, and G stands for the oscilla-

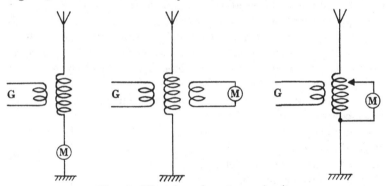

Fig. 98. Microphone in antenna circuit.

tion generator of whatever kind. The resistance of the antenna circuit, and therefore the amplitude of the high-frequency current in it, then fluctuates in accord with the voice. The maximum power variation in the antenna cannot then very much exceed the mean power consumed as heat within the microphone; but as long as the latter does not exceed a couple of watts or so, excellent results are obtainable in this way with the ordinary microphones of line telephony. A slight modification consists in placing the microphone in an intermediate circuit between the generator and the antenna, as in Fig. 99; but the same limitation of power obtains.

When the microphone is traversed by the high-frequency current. it must, of course, be situated very near to the antenna; and this may be inconvenient or impossible in some circumstances.

To deal with larger powers, un- usual microphone arrangements must be employed. The Poulsen arc was for long available as a practicable means of producing relatively large antenna powers, and the only bar to long-distance wireless telephony was the absence of any microphone capable of con-

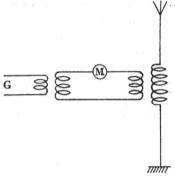

Fig. 99. Microphone in inter- mediate H.F. circuit.

trolling large power with good articulation. Great efforts were made by many inventors to produce satisfactory groups of the ordinary small carbon-granule microphones; and various types of microphone were devised not employing carbon contacts at all. But so little success attended these efforts that it may be doubted whether long-distance wireless telephony would ever have become practicable if it had continued to depend on control by means of a high-power microphone.

With the advent of the high-frequency alternator, a form of relay-control became possible, at least in theory. The output from the alternator is entirely de- pendent on the field magnets, so that a microphonic control of the field would control the ampli- tude of the antenna oscillation. This method is illustrated dia- grammatically in Fig. 100, where F is the field magnet of a high- frequency alternator and M the microphone. Unfortunately, in a small alternator the power used

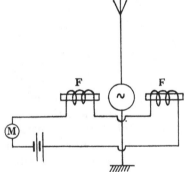

Fig. 100. Microphone in field circuit of alternator.

to excite the field cannot be made very small compared with the high-frequency output of the machine; so that with the ordinary small microphone no very great advance is made. In any case, the high-frequency alternator is adapted rather to large powers and very long waves than to small powers and short waves.

An entirely different method of control, of great interest, is based upon the variation of the permeability of iron with the magnetic induction. The principle appears to have originated with L. Kuhn* of the Telefunken Company, who used the unidirectional magnetising current in an auxiliary winding to vary the inductance (for high-frequency current) of the main winding around the same iron magnetic circuit. An improved form is due to E. F. W. Alexanderson, and is known as his "magnetic amplifier." Its principle is shown in Fig. 101. The antenna is excited by a C.W. generator G of any type. C_1, C_2, C_3, C_4 are the four cores of a peculiar iron-cored inductance suitable for high-frequency currents, and constructed therefore of very finely laminated Stalloy (or other magnetic alloy of high electric resistivity). The cores C_2 and C_3 form the magnetic circuit of the two coils, one on each core, inserted in the antenna circuit. The inductance of these coils is therefore proportional to the permeability in C_2 and C_3. C_2 and C_3 carry a second

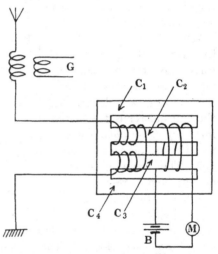

Fig. 101. Magnetic amplifier.

winding, wound over both cores together, the magnetic circuit of which consists of $(C_2 + C_3)$ and $(C_1 + C_4)$ as go and return paths respectively. In this winding flows a current from the battery B, modulated by the microphone M. The magnetic induction, and therefore the permeability, in C_2 and C_3 are thus controlled by M; consequently the natural frequency of the antenna is a function of the current in the microphone. The object of the four-core arrangement is the avoidance of high-frequency E.M.F. in the microphone circuit.

In the arrangement shown in Fig. 101, the action of the micro-

* *Elektrot. Zeitschrift*, 1914; or *Jahrb. d. dr. Telegraphie*, June 1915.

phone is to vary the amplitude of antenna current by distuning
the antenna from the generator **G**. It is obvious that the magnetic
amplifier may be similarly used to distune the generator itself,
so causing a fluctuation of
radiated wavelength (with or
without considerable change
of amplitude) under the action
of speech in the microphone.
With a magnetic amplifier
used in this way in association with the Alexanderson
alternator at New Brunswick,
speech has been transmitted
as far as 3,000 miles.

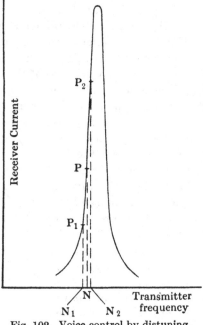

Fig. 102. Voice control by distuning.

When speech control takes
the form of variation of
wavelength rather than amplitude at the transmitter,
the varying wavelength of
the transmitter is translated
into varying amplitude at the
receiver as indicated in Fig.
102. The receiving circuits
are adjusted to resonate at a
frequency differing slightly
from the no-speech transmitter frequency N, thus bringing the
representative point on the resonance curve to some position P on
one of its steep sides. When speech occurs, the transmitter frequency varies between N_1 and N_2, and the receiver current amplitude
varies between N_1P_1 and N_2P_2.

3. THE AID OF THE TRIODE IN MICROPHONIC CONTROL

The use of the triode, singly or in groups, extends almost indefinitely the possibilities of controlling large antenna power by
means of a feeble microphone. Firstly, its mere amplifying properties can be applied to reinforce the action of the microphone in
the arrangements of Figs. 100 and 101. The microphone can be
made to modulate the potential of the grid, where the absorption
of power is small or zero, and the thereby controlled anode current
be made to flow in the field winding of the alternator (Fig. 100)

or the control winding of the magnetic amplifier (Fig. 101). The latter, for instance, then develops, in its simplest form, into the arrangement of Fig. 103. If the grid is kept at a negative potential, the only current taken from the secondary of the microphone transformer is due to stray capacities; so that the microphone imposes no limit on the power controllable by the anode of the tube. A very large triode, or a large group of triodes in parallel, can then be controlled by the ordinary small microphone of line telephony. Moreover, cascade amplification can be resorted to, any type of low-frequency amplifier* being interposed between the microphone and the triode.

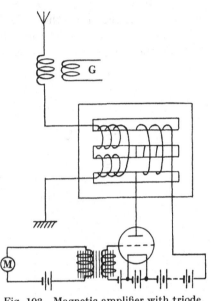

Fig. 103. Magnetic amplifier with triode.

Secondly, the damping effect of a triode with its anode-filament connected across the whole or part of the inductance of an oscillatory circuit may be utilised. In Fig. 104, the *LC* circuit is damped by the presence of the triode; and the decrement introduced by the triode is the greater as that proportion of the cycle is increased over which the anode current is neither zero nor saturated. By arranging that this proportion shall vary with the grid potential, a microphone which influences the latter is made to influence the amplitude of the high-frequency oscillation in *LC*.

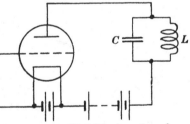

Fig. 104. Variable damping of oscillatory circuit.

One way of accomplishing this is depicted in Fig. 105. Here

* See Chapter VII, Section 3.

two of a family of i_a, v_a characteristics for constant values of v_g are shown, one for $v_g = V$ and another for a lower grid potential $v_g = V'$. When $v_g = V$, let OM = ON be the amplitude of high-frequency E.M.F. from the oscillatory circuit applied between anode and filament. During the negative half-cycle (corresponding to ON), the triode takes no energy from the oscillatory circuit;

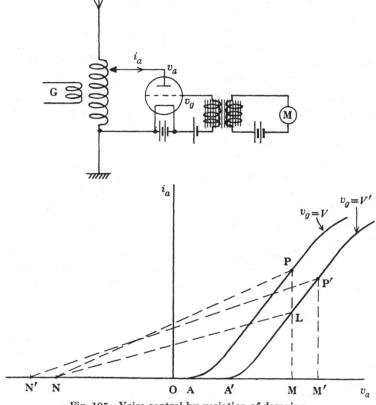

Fig. 105. Voice control by variation of damping.

but during a large part (corresponding to AM) of the positive half-cycle, the triode does absorb energy from the oscillatory circuit. Its average damping effect is that which would be produced by a conductance approximately equal to the average slope $\dfrac{di_a}{dv_a}$ of the curve (say approximately the slope of the line NP). Now suppose that v_g fluctuates at an acoustic rate owing to speech at the microphone. When the E.M.F. in the secondary of the microphone transformer is such as to change v_g to the value V', the triode absorbs energy from the oscillatory circuit during a much smaller

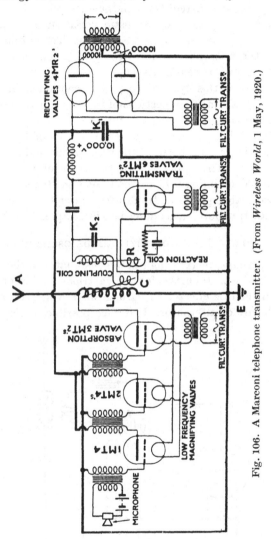

Fig. 106. A Marconi telephone transmitter. (From *Wireless World*, 1 May, 1920.)

portion of the positive half-cycle (viz. the portion corresponding to A'M instead of AM). The decrement of the oscillatory circuit being thereby reduced, the oscillation increases until equilibrium

is reached at some amplitude OM' = ON'; the triode is then responsible for as much damping as would be produced by a conductance approximately equal to the slope of the line N'P'. An
acoustically varying current through the microphone thus produces
a more or less proportionate acoustic variation of amplitude of
the high-frequency oscillation in the antenna. The antenna current
rises when the microphone causes the grid potential to fall, and
falls when the grid potential rises.

An example of this type of control is contained in Fig. 106,
which is the diagram of connections of the transmitter used by the
Marconi Company at Chelmsford during some telephony experiments with Madrid. There are here two stages of low-frequency
amplification interposed between the microphone and the damping
group of triodes. The apparatus to the right of the antenna is
the unspecified oscillation generator G of Fig. 104, and is, in this
case, an oscillating group of triodes fed through a pair of valves
from an A.C. transformer, as described in Chapter IX (Fig. 91).

4. SPEECH CONTROL WITH TRIODE OSCILLATOR

Except, perhaps, in very high-power installations, the generator
is likely always to take the form of a triode oscillator. Speech
control can then be effected in various ways which are not applicable to the arc or alternator. It is even possible to dispense with

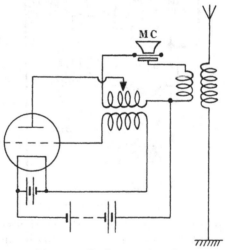

Fig. 107. Condenser microphone.

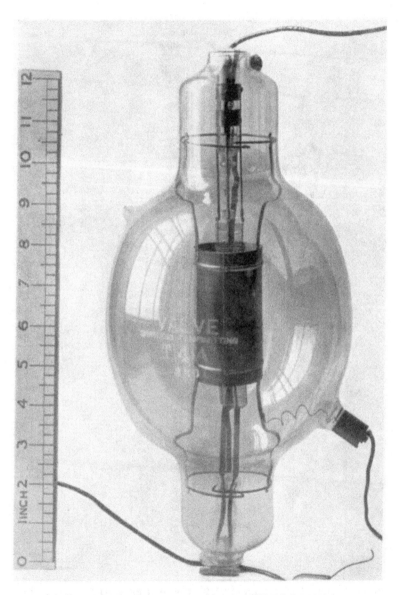

PLATE XXII. TRANSMITTING TRIODE "T4A" (p. 128).

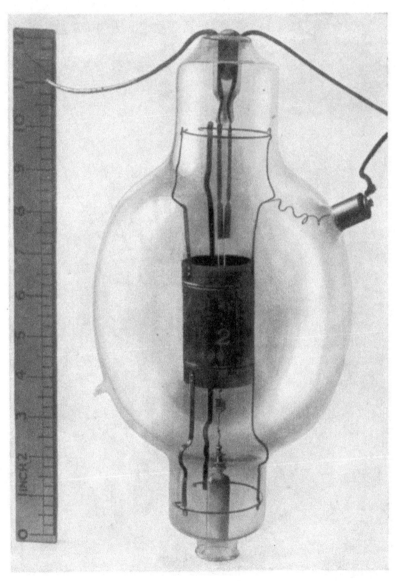

PLATE XXIII. VALVE (DIODE) "U2" FOR RECTIFYING H.T. ALTERNATING CURRENT (p. 140).

the crazy-contact microphone altogether, replacing it by some device whose capacity or inductance varies under the action of the voice. Thus in Fig. 107, MC indicates diagrammatically a microphonic condenser, one of whose plates is in the form of a diaphragm displaceable by sound waves impinging on it through the mouthpiece. The voice thus controls the capacity of the condenser and the frequency of the oscillation generated by the triode, and consequently the wavelength (and amplitude) of the radiation.

The principle of one of the commonest speech control circuits, with which excellent articulation is easily obtained, is shown in Fig. 108. We have seen (Chapter IX, Section 5) that in the oscilla

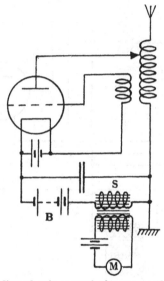

Fig. 108. Microphonic control of mean anode potential.

ting triode stability of amplitude is reached either through lack of anode current or through lack of anode potential. The oscillator in Fig. 108 is adjusted so that growth ceases from the latter inhibition; any rise or fall of the voltage of the anode battery B would then be accompanied by a nearly proportionate rise or fall of oscillatory current. Superposed on the battery E.M.F. in the anode supply is the E.M.F. in the secondary S of the microphone transformer. The amplitude of the high-frequency current is therefore varied in proportion to the fluctuation of current in the microphone, provided that the speech E.M.F. in S is never great enough to make the anode current reach zero or saturation value.

The arrangement of Fig. 108 is, of course, just as limited with respect to power controllable by a feeble microphone as is that of Fig. 98. But the power controlled by the same feeble microphone can be increased without theoretical limit in either or both of two ways. Firstly, a high-frequency amplifier can be inserted between the weak speech-controlled oscillator and the antenna, as indicated in Fig. 109. Secondly, a low-frequency amplifier can be inserted

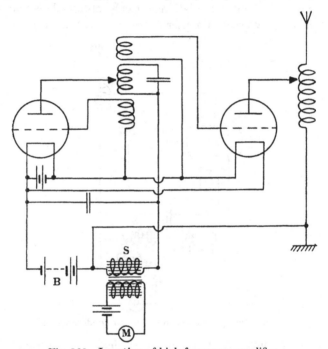

Fig. 109. Insertion of high-frequency amplifier.

between M and S, precisely as in Figs. 103 and 106. In its simplest form, this arrangement is shown in Fig. 110. S is now attached to the grid of the control triode CT, its place in the anode supply circuit of the oscillating triode OT being taken by a choke L through which passes the anode current of both triodes. It is obvious that whenever the grid of CT is rising (say) in potential, the anode current of CT (flowing in L) will be rising, and the voltage of the supply to OT is reduced by the corresponding reactance drop in L.

This form of wireless telephone transmitter is of great practical importance*, the method of speech control being known as that of the "anode choke." The next Section contains a cursory investigation of the conditions governing the operation of such circuits

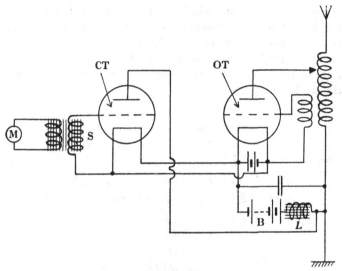

Fig. 110. Insertion of low-frequency amplifier.

5. ANALYSIS OF ANODE CHOKE CONTROL

The oscillator itself (OT and the antenna in Fig. 110) has already been analysed in Chapter IX; but we here enquire what should be the dimensions of the choke L and the triode CT by which speech control is effected, and what relation obtains between the power of the oscillator and the power dissipated in the control triode.

Let an acoustic sinoidal E.M.F., represented by the rotating vector E, occur in the secondary winding S; and let the consequent acoustic fluctuations of currents be represented by the vectors

I_c in anode of CT,

I_o in anode of OT (mean value as regards the high frequency),

and therefore $(I_c + I_o)$ in choke L.

---

* It is used, for example, in the R.A.F. wireless telephone air and ground stations. See C. E. Prince, "Wireless Telephony on Aeroplanes," *Journ. Inst. of Elect. Engnrs.*, May 1920.

The acoustically fluctuating part of the high-tension supply to the oscillating triode OT (and incidentally to CT as well) is then the vector

$$- jpL \, (I_c + I_o)$$

where j is written for the operator $\sqrt{-1}$. It is this fluctuation in the anode supply voltage which produces a proportionate fluctuation of high-frequency current; and in order to arrive at sensitive control, the designer wishes to make the ratio between it and the microphone E.M.F. E as large as possible.

If a_o is the apparent anode conductance of the oscillating triode, we have

$$I_o = - jpL \, (I_c + I_o) \, a_o$$

$$\therefore \; I_c = - \left(\frac{1 + ja_o pL}{ja_o pL} \right) I_o = - \left(\frac{a_o pL - j}{a_o pL} \right) I_o.$$

And for the control triode

$$I_c = a_c \, [- jpL \, (I_c + I_o)] + g_c E$$

$$\therefore \; (I_c + I_o) = - jpL \, (I_c + I_o) \, (a_c + a_o) + g_c E$$

$$\therefore \; \frac{jpL \, (I_c + I_o)}{E} = \frac{jpLg_c}{1 + jpL \, (a_c + a_o)}$$

$$= \frac{pLg_c \, [pL \, (a_c + a_o) + j]}{1 + p^2 L^2 \, (a_c + a_o)^2}.$$

The magnitude of this ratio, which we may call the control sensitivity, is

$$\frac{pLg_c}{\sqrt{1 + p^2 L^2 (a_c + a_o)^2}}.$$

For choice of L, the control sensitivity is greatest when the reactance pL is made indefinitely great compared with $\dfrac{1}{a_c + a_o}$; it is then

$$\frac{g_c}{a_c + a_o}$$

$$= \frac{\nu_c}{1 + \dfrac{a_o}{a_c}}$$

where ν_c is the amplification factor $\dfrac{g_c}{a_c}$ of the control triode.

With a given ν_c*, therefore, for sensitive control a_c should be large compared with a_o. Under the best conditions, when

---

* Limited in practice by the available high-tension supply, as explained in Chapter VII, Section 2.

$a_c > > a_o$, the fluctuation of supply voltage to the oscillating triode reaches nearly v_c times the microphonic fluctuation of grid potential at the control triode.

With regard to the dissipation of power in the control triode, we have seen that

$$I_c = - \left(\frac{a_o pL - j}{a_o pL} \right) I_o.$$

Hence when $pL > > \dfrac{1}{a_o}$, the magnitude of I_c is approximately equal to that of I_o, and these currents are approximately in anti-phase. Speech in the microphone then produces no change in the current from battery B; and the function of CT is to make L alternately divert energy from, and restore this energy to, the oscillator. This, however, does not imply that there is no waste of power in CT. If the oscillation in OT is fully modulated by the speech, the magnitude of the fluctuation I_o must be equal to the mean anode current in OT—i.e. the mean anode current in OT must vary during speech between zero and twice the no-speech value. The magnitude of I_c being the same as that of I_o, it follows that CT requires as large an anode current as that of OT.

The foregoing analysis will enable the designer to calculate the dimensions of the control circuit required for modulating any given oscillator. Incidentally it should remove the misconception* that the power dissipated in the control triode can be made negligible in comparison with the power of the oscillator. If there is to be full speech modulation, of the power supplied by the high-tension battery B not more than half can be delivered to the oscillator triode and at least half must be dissipated in the control triode.

6. DUPLEX WIRELESS TELEPHONY

With Morse signalling, the only object in duplexing† the con-nection is to increase the total speed of working between the two stations. In telephony, on the other hand, there is a much more urgent need for duplex working, as anyone who has tried to chat through a speaking tube will recognise. The necessity for each of the two performers to exchange the rôles of speaker and listener by mutual agreement renders easy conversation impossible.

---

* See discussion of C. E. Prince's paper, *loc. cit.* p. 163.

† A "simplex" connection is one along which communication can pass in only one direction at a time. A "duplex" connection carries communication in both directions simultaneously. The ordinary line telephone provides a duplex connection.

Under ordinary conditions a wireless receiver is put completely out of action by any transmitting on the same or a near antenna, even though different wavelengths are used for sending and receiving. This difficulty has been met in telegraphy, when duplex working has been provided, by locating the sending and receiving stations many miles apart, and employing directive antennae and balancing devices to eliminate the transmitted signals from the receiving apparatus. The same devices can be resorted to for telephony, but less conveniently because the sending and receiving must here be performed by a single person. Moreover, such an arrangement would add to the difficulties of connecting the wireless telephone to the ordinary land-line system. Some means are therefore needed by which sending and receiving can be performed in a single station by a single untrained person.

This problem is receiving the attention of inventors, and numerous solutions have been suggested. They may be classified broadly as (a) devices in which the screening of the receiving apparatus from the transmitter is carried to the necessary extreme pitch of refinement; and (b) devices in which the transmitter radiates only during the moments of actual speech. The former depend on various methods of obtaining directive discrimination, on screening, balancing and tuning out: they will not be further discussed here*. The latter do not strictly give duplex service at all, since speaking and hearing cannot be practised simultaneously. They do not enable the listener to "break in" upon the speaker; but they do free the speaker from performing any switching operation or even advising the listener that their rôles are now to be reversed. Devices of class (b), therefore, do go a fair way towards meeting the difficulty.

In Fig. 108, if the battery B is omitted, or is reduced in voltage so as to be insufficient to set the triode oscillating, oscillation will occur only at times when there exists in S an E.M.F. of suitable sense and sufficient strength. There is then no oscillation in the antenna except while the microphone is being agitated by the voice; and then the more violent the agitation the stronger the oscillation. Only one-half of the acoustic cycle is utilised, but this is no serious objection; and in any case by providing two oscillator triodes it would be possible to make both half-cycles active. The

* See, however, Chapter XII, Section 4, (b).

same result is achieved in the arrangement of Fig. 110 by inter-
changing L and B, so as to feed CT from the battery and OT
mainly or wholly from the choke, as shown in Fig. 111.

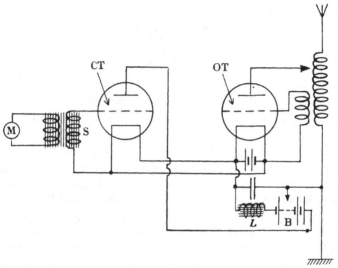

Fig. 111. Transmitter oscillating only during speech in microphone.

In practice, serious difficulties are met in endeavouring to realise
arrangements of this type. It is easy, without loss of articulation,
to make the oscillation very feeble in the absence of speech; but
if there is then absolutely no oscillation, very great difficulty is
experienced in retaining good articulation. Any perceptible
threshold effect seems to be fatal. It is possible that these difficulties may be overcome; but if they are not, the method may be
of use as an auxiliary which greatly facilitates further protection
of the receiver by the various devices of class (a).

CHAPTER XII

MISCELLANEOUS

1. Transmitting antennae

ANTENNA has been seen in Chapters II and III to be the name given to an oscillatory circuit whose radiating (and therefore, conversely, absorbing) property is strongly marked. In its earliest form, the Hertz oscillator, it consisted of a straight wire, with or without plates or balls at the ends, broken by a spark gap at the middle. There was therefore an antinode of current in the middle and nodes at the ends; and one end rose to its maximum positive potential when the other end reached its maximum negative potential, as in Fig. 112 A. In the case of the plain wire, the wavelength of the

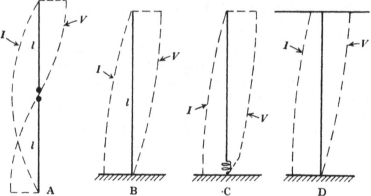

Fig. 112. Distribution of current and potential along antennae.

radiation was then about four times the length l of the half oscilla-tor*. G. Marconi's vital contribution from an engineering stand-point to the early scientific work of H. R. Hertz and O. J. Lodge was the enlargement of the short Hertz oscillator into an extensive

* The exact figure has been the subject of much mathematical discussion. It appears to be about 4·2.

structure supported high above the ground on masts. He also dispensed with the lower half of the oscillator by joining the current antinode to earth, thus arriving at the simple vertical antenna of Fig. 112 B, with a wavelength (unloaded) of about 4·2 times the length l, and a distribution of current amplitude falling approximately sinoidally along the wire from a maximum at the bottom to zero at the top. Loading the antenna by an inductance inserted at the bottom lengthens the wave, and modifies the amplitude distribution as in Fig. 112 C.

At any given frequency, the contribution of each part of the wire towards the total power radiated is proportional to the square of the current amplitude in it. If, therefore, all the current at the base of the antenna could be made to flow all the way up to the top, for the same base current the power radiated would be increased; the antenna would have a larger radiation resistance*, and the higher the antenna the larger this radiation resistance would be. This is the primary object in attaching to the top of the vertical wire a more or less horizontal system of large capacity to earth. The distributions of current and potential amplitudes in the vertical wire then become those of Fig. 112 D. When the capacity at the top has been made so large compared with the effective capacity of the uplead that the current in the latter is sensibly uniform, the length of that current is sensibly the full height h of the uplead, which is the greatest available height with the given masts. For low-power transmitters there is then no advantage in further increasing the capacity.

With unchanged antenna and wavelength, an increase of power radiated implies larger current in the uplead, and therefore larger CV in the top, C being the capacity of the top and V its potential amplitude. When such increase brings V anywhere near the potential for brush discharge—say 50,000 to 100,000 volts—it is necessary to enlarge C still more. With given height of antenna, large power demands, therefore, an extensive, more or less hori-

---

* See Chapter III, Section 6. The radiation resistance r_r is that part of the total resistance which is due to the radiation. If $\mathscr{J}$ is the R.M.S. current at the earth connection, $\mathscr{J}^2 r_r =$ power radiated. In an antenna with large flat top at height h, $r_r = 1{,}600 \dfrac{h^2}{\lambda^2}$ ohms. In a vertical loop antenna of area S and number of turns T, $r_r = \dfrac{63{,}000 S^2 T^2}{\lambda^4}$ ohms. (M. Abraham, *Jahrb. d. dr. Teleg.*, Aug. 1919.)

zontal top carried on several masts*. Since it is the peak antenna potential which must not reach the brushing value, the need for large antenna capacity is more pronounced in spark than in C.W. transmitters of equal power.

We have treated the wavelength as specified without reference to height and capacity of the antenna. Although the minimum (i.e. unloaded) wavelength is, of course, dependent on height and capacity, the existence of an optimum wavelength for propagation efficiency (Chapter II, Section 3), or other considerations, usually calls for waves longer than the minimum. Hence, briefly, the capacity of a transmitting antenna (and therefore also the number of masts) is made large firstly to make the effective height of the antenna approximate to the height of the masts, and secondly in order to restrict the peak voltage reached; and the height of the masts is made large to obtain an antenna of large radiation resistance.

What, then, is the significance of the radiation resistance of a transmitting antenna? The power radiated is $\mathscr{I}^2 r_r$, where $\mathscr{I}$ is the R.M.S. current and r_r the radiation resistance of the antenna. Now the antenna circuit, like every other circuit practically realisable, loses energy also by conversion into heat. Hence there is a loss of power $\mathscr{I}^2 r_h$, where r_h may be called the heat resistance of the antenna. The total resistance is then

$$r_t = r_r + r_h.$$

Of the high-frequency power supplied to it from the generator, the antenna radiates $\mathscr{I}^2 r_r$ and wastes $\mathscr{I}^2 r_h$. Its efficiency is $\dfrac{r_r}{r_t}$, and this ratio must be made as large as possible. There is no theoretical limit short of unity.

Although a single vertical wire may have a radiation resistance up to about 36 ohms (its radiation resistance when oscillating at its unloaded wavelength), high masts are such costly structures† that except with very short waves it is out of the question to provide antennae with anything like such high radiation resistances.

---

* At Čarnarvon there are 10 masts 400 feet high supporting an antenna ⅔ mile long and 500 feet wide. The capacity is about ·04 microfarad. At Hanover there is a central mast 820 feet high and six masts 400 feet high supporting an umbrella antenna. At Nauen, where the antenna is said to receive 400 kilowatts, it is carried on two masts 850 feet high and covers a ground area of 40 acres.

† In a high-power station the masts may cost nearly as much as all the rest of the station.

A small fraction of an ohm is an ordinary value. Moreover, although the metallic resistance of the antenna wires and inductance coils can be kept small, large heating losses occur in the poorly conducting soil at the connection between antenna and earth, and in the soil below the antenna. In large stations these losses are minimised by the use of extensive systems of conductors buried a few inches in the soil; but even so it is generally found impossible to bring r_h below several ohms*. The efficiency of the antenna is then very low. Thus for the large station at New Brunswick, N.J., E. F. W. Alexanderson states† that at $\lambda = 13,600$ metres, the radiation resistance was ·07 ohm and the total resistance 3·8 ohms, giving an antenna efficiency of less than 2 per cent.

In engineering practice, wherever efficiencies are low, hopes for improvement may be high. By substituting a new form of antenna with multiple earth connections—in effect, a number of separate antennae connected in parallel—without changing the radiation resistance Alexanderson has succeeded at New Brunswick in reducing the total resistance from 3·8 ohms to ·5 ohm, thus raising the antenna efficiency to about 12 per cent. At a shorter wave, 8,000 metres, the radiation resistance with this arrangement was ·2 ohm and the total resistance ·6 ohm, giving the remarkable antenna efficiency of 25 per cent. Striking improvements in long-range transmitting stations are to be expected from developments in this direction‡.

2. RECEIVING ANTENNAE

For reception, the conditions governing antenna design are very different. To obtain the largest possible antenna efficiency§ it is still necessary to use the highest possible antenna. By raising the height of the antenna we increase the E.M.F. produced in it by the incoming signal without increasing its heat resistance; and we thus increase that fraction of the power transmitted which becomes available for the detector. There are, however, two distinct reasons why such increase of E.M.F. is often of no practical advantage.

* Even in ship stations, where sea water replaces soil, r_h cannot be reduced below 2 ohms or so.

† "Trans-oceanic radio communication," *Proc. Am. Inst. Elect. Engrs.*, Oct. 1919.

‡ See two important papers by G. W. O. Howe: "The efficiency of aerials," and "The power required for long distance transmission," *Radio Review*, Aug. and Sept. 1920.

§ That is, "intrinsic efficiency"; see footnote on p. 173 below.

Firstly, reception often fails not directly from weakness of the incoming signal but from the presence of relatively strong interfering atmospherics. As in the parallel case of parasitic disturbances in low-frequency amplifiers (Chapter VII, Section 4), nothing is gained by increasing together the signals and the disturbances. The almost unlimited sensitiveness of modern triode amplifiers strongly reinforces this consideration, and one may consider that reception nowadays always fails by interferences of one sort or another rather than by intrinsic weakness of the incoming signal. The need, therefore, is for further discrimination rather than for the increase of receiver sensitivity which would be given by elevating the antenna.

Secondly, the use of retroactive triode circuits makes it possible to reduce the resistance of any circuit to any extent (Chapter X, Section 1); and this profoundly modifies the conditions governing the design of the receiving antenna. The formula on page 15 for antenna currents shows that the current in the receiving antenna is proportional to its height and therefore to the square root of its radiation resistance (see footnote on page 169); and to the reciprocal of its total resistance. Hence

$$\mathscr{J} = k \cdot \frac{\sqrt{r_r}}{r_t}$$

where k is a constant while the transmitting conditions remain unchanged. The total resistance r_t in the receiving antenna is the sum of the radiation resistance r_r, the heat resistance r_h, and what may be called the detector resistance r_d corresponding with the power taken from the antenna by the detector circuits coupled to it. The useful power, viz. the rate of working on the detector circuits, is therefore
$$\mathscr{J}^2 \cdot r_d = \frac{k^2 \cdot r_r \cdot r_d}{r_t^2}.$$

This is a maximum for choice of r_d (which is entirely under control) when $r_d = \frac{1}{2} r_t$, and is then

$$\frac{k^2 r_r}{2 r_t}.$$

By triode retroaction r_t can be made as small as desired: let it be reduced to a fraction $\frac{1}{m}$ of the radiation resistance r_r. The detector power is then $m \frac{k^2}{2}$, and is independent of the height of the antenna.

The same result—detector power independent of height—is found for a loop antenna. Hence any two antennae of any sorts or sizes for which the ratios $\frac{r_r}{r_t}$ are equal are equally effective in delivering power to the detector*.

This remarkable fact not only has enabled us to dispense with very high receiving antennae, with their costly masts, but has even led to the use of relatively very small closed loops of extremely low radiation resistance, on account of incidental advantages conferred by their directive and other qualities. Already a large proportion of long-range wireless traffic is received on loop antennae in the form of a coil of several turns of wire carried on a frame or on small masts. The shape is usually rectangular, and the dimensions commonly lie between (say) 10 feet square and 100 feet long by 30 feet high. As an extreme example of what may be done with diminutive antennae by the use of triodes, it may be mentioned that the signals from Annapolis, U.S.A., have been received in Paris with an antenna consisting of a coil of wire 10 inches in diameter.

There are, however, two practical limitations preventing us from pushing this principle indefinitely far. One is the difficulty of maintaining the extremely steady retroactive adjustments necessary for keeping the total antenna resistance almost zero but not negative. The other is that the time required to reach any specified useful fraction of the final steady amplitude of oscillation (which is approached asymptotically) increases as the resistance is decreased, and may become so great as to restrict the speed of signalling. Both these limitations have been referred to in Chapter x, Section 1.

* Since detector power is a maximum when $r_d = \frac{1}{2}r_t$, the ratio

$$\frac{\text{transmitter power delivered to detector circuits}}{\text{transmitter power reaching receiving antenna}},$$

which may be called the intrinsic efficiency of the receiving antenna, cannot exceed 50 per cent. But with retroactive arrangements power derived from a local (anode) battery but controlled by the signal is delivered to the detector. The ratio

$$\frac{\text{total power delivered to detector circuits}}{\text{transmitter power reaching receiving antenna}},$$

which may be called the total efficiency of the receiving antenna, may then be given any value up to infinity.

3. WIRELESS DIRECTION FINDING

Any antenna symmetrical about a vertical axis, such as a single vertical wire or an umbrella antenna, obviously radiates (and absorbs) equally in all horizontal directions. Even asymmetric open antennae, such as an inclined single wire, an inverted L or a T antennae, show very slight directive effects. An extreme case of the inverted L, however, in which the horizontal top is very much longer than the vertical uplead, does show well-marked directive properties, and radiates most intensely in the direction tip-to-uplead*. This type has been much used by the Marconi Co. as an aid in obtaining duplex working with separated transmitting and receiving stations. An example is the antenna at Carnarvon (see footnote on page 170), where the length of the horizontal top is about nine times the height.

Calculable and very marked directive effects are obtainable by associating two or more antennae arranged to oscillate isochronously with definite phase relations. A particularly simple and interesting case is that shown in Fig. 113. AA′ and BB′ are two vertical

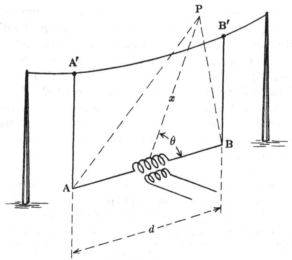

Fig. 113. Directive antenna.

* The theory of its directive action is a matter of some controversy, and appears to depend on the imperfect conductivity of the ground beneath. There is evidence, too, that the unsymmetrical distribution observed fairly near to the transmitting antenna no longer obtains at great distances—a phenomenon probably common to all directive transmitters.

wires hung from insulators at A′ and B′, distant d apart, containing oscillations of the same amplitude and with phase difference 180°. P is a remote point distant x from the antenna in a horizontal direction inclined at θ to the plane of the antennae. At any instant the electric or magnetic field at P is the sum of the fields due to the two radiators at A and B. Since the radiators are in anti-phase, the components at P differ in phase by

$$\pi - \frac{PA - PB}{\lambda} \cdot 2\pi$$

$$= \pi \left(1 - \frac{d}{\frac{1}{2}\lambda} \cos \theta\right) \text{ since } x > > d.$$

If a is the amplitude of each component, the resultant amplitude at P is

$$2a \cos \frac{\pi}{2} \left(1 - \frac{d}{\frac{1}{2}\lambda} \cos \theta\right)$$

$$= 2a \sin \left(\frac{\pi}{2} \cdot \frac{d}{\frac{1}{2}\lambda} \cdot \cos \theta\right).$$

Assuming that $d \not> \frac{1}{2}\lambda$, this varies from zero when $\theta = 90°$ to a maximum $2a \sin \left(\frac{\pi}{2} \cdot \frac{d}{\frac{1}{2}\lambda}\right)$ when $\theta = 0°$; it is then $2a$ if $d = \frac{1}{2}\lambda$, i.e. if the antennae are half a wavelength apart. The shape of the polar curve showing amplitude as a function of θ depends on the ratio $\frac{d}{\frac{1}{2}\lambda}$; and when this is small it approximates to a pair of equal tangent circles of diameter $2a \cdot \frac{\pi d}{\lambda}$. Fig. 114 shows the curve for the cases $\frac{d}{\frac{1}{2}\lambda} = \frac{1}{10}$ and $\frac{1}{2}$, the former being indistinguishable from a pair of circles.

In the converse case of the antenna receiving instead of transmitting, the same function of θ gives the amplitude of the E.M.F. produced in the antenna circuit by incident radiation which makes an angle θ with the plane of the antenna.

A vertical loop antenna possesses the same directive property, the vertical portions of the loop—or, if the loop is not rectangular, the vertical components of the two halves of the loop—behaving as the pair of vertical antennae just examined. A receiving antenna in the form of a loop rotatable about a vertical axis can accordingly be used to find the orientation of the incident radiation, the signals being strongest when the plane of the loop contains the rays and vanishing when the plane is perpendicular to the rays.

It is clear from Fig. 114 that when θ is small $\dfrac{dr}{d\theta}$ is small, so that the position of maximum intensity could not be precisely determined; but near the position of zero intensity, $\dfrac{dr}{d\theta}$ is very large

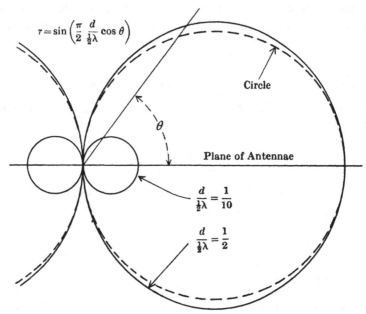

$$r = \sin\left(\frac{\pi}{2}\,\frac{d}{\frac{1}{2}\lambda}\,\cos\theta\right)$$

Circle

θ

Plane of Antennae

$$\frac{d}{\frac{1}{2}\lambda} = \frac{1}{10}$$

$$\frac{d}{\frac{1}{2}\lambda} = \frac{1}{2}$$

Fig. 114. Distribution of intensity around directive antenna.

and the zero position can be much more precisely determined. We have already seen that modern triode amplifiers make the loop a practicable form of receiving antenna; rotatable loops are accordingly much used as direction finders. By observing two positions of approximately equal intensity, one on each side of the position of zero or minimum intensity, the operator is able to determine the orientation (but not the direction) of the incoming rays with remarkable accuracy.

For mechanical reasons it is not practicable to use large rotatable loops, particularly when determinations have to be made rapidly. A development in principle of the rotatable loop method, though historically antecedent, is due to E. Bellini and A. Tosi, with improvements by the Marconi Co. In Fig. 115, A_1, A_2 are two equal loop antennae fixed at right angles; and being fixed, they may be

PLATE XXIV. TELEPHONE TRANSMITTER AND RECEIVER (MARCONI) (p. 152).
(The Wireless Press, Limited.)

large single-turn loops carried on masts instead of small multi-turn loops on movable frames. Each is connected to one of two mutually perpendicular and centrally situated primary windings P_1, P_2 of a "radiogoniometer," within which a central secondary coil S is rotatable. The two circuits $A_1P_1C_1$, $A_2P_2C_2$ are made precisely similar, and are set precisely at right angles so that they have no mutual inductance. If each is exactly tuned to the incoming signal (or equally distuned) by means of the condensers C_1, C_2, the oscillations produced will be in phase; and if the width of the loops is small compared with λ, the amplitudes will be proportional to $\cos\theta$ and $\cos\left(\dfrac{\pi}{2}-\theta\right)$, where θ is the inclination of the incident rays with the plane of one of the antennae (say A_1). The two E.M.Fs. induced in the secondary S by reason of its mutual inductances with P_1 and P_2 will balance for the two positions of S given by

$$\cos\theta\cos\phi + \cos\left(\frac{\pi}{2}-\theta\right)\cos\left(\frac{\pi}{2}-\phi\right) = 0$$

where ϕ is the angle between S and P_1;

i.e. $\tan\theta = -\cot\phi$

$$\theta = \phi \pm \frac{\pi}{2}.$$

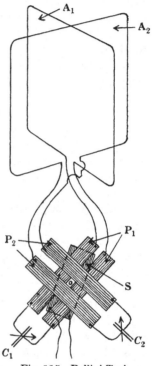

Fig. 115. Bellini-Tosi arrangement.

Thus by reading on a scale the position ϕ for zero signals, the orientation (but again not the direction) of the incident radiation is found.

By either of these methods determinations to within 1° are practicable. In both, various precautions are necessary if a good— i.e. nearly zero—minimum strength is to be obtained. These precautions are mainly concerned with avoiding effects due to the loop antennae acting partly as open antennae by virtue of their capacities to earth. In the Bellini-Tosi system there is the added practical complication with spark signals of tuning the two antennae to be so precisely alike that the two oscillations not only have sensibly equal amplitudes but are in sensibly exact synchronism; for without synchronism a rotating field is produced in the goniometer and no position of S will give balance. As it is impossible to carry out the synchronising adjustment rapidly, the two antennae are generally tuned accurately together at a wavelength near that of the signals to be received, and so left. Variation of the incoming wavelength then varies the strength of the signal but does not destroy the minimum. Indeed the antennae are sometimes rendered actually aperiodic by short-circuiting the condensers C_1, C_2. They become then much less sensitive, but the synchronising adjustments are avoided.

Even when finding directions by a single station, the ambiguity of 180° in the determination is seldom much inconvenience, as the direction of the source of signals is generally known to within 90°. Direction-finding stations on land, however, are usually grouped in pairs at the ends of a suitable base line, so that the intersection of the two orientations determined gives the position of the source of signal, e.g. a ship or an aeroplane.

Since it is the orientation of the rays at the receiving station which is determined, the direction of the sending station can only be inferred on the assumption that the rays are straight (or, more strictly, lie in vertical planes). This is certainly not always the case. Large departures, sometimes varying spasmodically by as much as 10° or 20°, are frequently observed at night. At night too, it is sometimes impossible to obtain a balance at all, indicating that two or more rays have reached the receiving station along paths of unequal length. During the daytime, these aberrations are much less prominent, both in magnitude and in frequency of occurrence.

4. INTERFERENCE FROM ATMOSPHERICS

Atmospherics, strays or X-s, as they are variously called, are the bane of wireless telegraphy. They make themselves felt as sharp clicks or crashing or grinding noises in the receiving telephones, and are often violent enough to drown altogether signals which in their absence would be called very strong. They are supposed to originate in electric discharges in the earth's atmosphere—e.g. lightning flashes—often at great distances from the disturbed station; and there may also be extra-terrestrial sources.

They are much more severe at some times than at others, summer generally being worse than winter and night than day. They vary enormously as between places on the earth, and are often particularly severe in the tropics, e.g. India. Improvement in our means for overcoming interference from atmospherics is the most pressing need in the wireless art to-day. With auditive reception the ear of the operator is able to exercise a good deal of discrimination when atmospherics are mild; but for high-speed telegraphy, where auditive reception is no longer possible, the elimination of atmospheric interference is especially necessary.

The solution of the problem obviously requires some means of discriminating at the receiver between the atmospherics and the desired signals. The transmitted signals might, of course, be given some special character which would aid in the discrimination*; but efforts hitherto have been mainly directed towards discriminating between atmospherics and the ordinary form of signals as received from spark or C.W. transmitters. Many suggestions have been made, in some cases leading to very considerable improvement. The possible methods of discrimination can be roughly classified as follows:

(a) By difference of frequency and/or decrement;
(b) By difference of direction of incidence;
(c) By difference of strength (amplitude);
(d) (a) with balancing devices; and
(e) (a) with amplitude limiting devices.

(a) The ordinary tuning arrangements by which one signal is distinguished from a simultaneous signal of different wavelength serve to favour the desired signal at the expense of the atmospheric. If the signal is from a C.W. transmitter, and the receiving circuits have very low decrement, it is feasible to use very loose couplings and so filter out the atmospherics a good deal. For example, in Fig. 52 arrangement C would be less troubled by atmospherics than arrangements A or B. But the atmospheric produces in the antenna an E.M.F. which, although it appears to be very highly damped or even aperiodic, may have an amplitude exceeding that

---

* It has been suggested, for example, to transmit C.W. radiation of frequency N, modulated in amplitude at a hyper-acoustic frequency n. The signal would be rendered audible at the receiving station by heterodyning the rectified signals with a local oscillation of frequency n', giving an audible beat of frequency $(n - n')$—instead of heterodyning the radio-frequency N, as usual.

of the signal hundreds or thousands of times. The antenna and other receiving circuits are themselves shocked by the atmospheric into oscillation at their own natural frequency; so that mere loose coupling and fine tuning can only handicap the atmospheric, which must win through if only it is strong enough.

(b) When disturbing signals come from one particular bearing provided that the desired signals do not emanate from nearly the same bearing, a loop antenna may be used turned broadside on to the origin of the disturbance. With a single source of disturbance, such as a powerful transmitting station not very far away, this procedure would serve, and is used, to cut out the interfering signals. Observations with a direction-finding receiver often show that the worst atmospherics come from a more or less well-defined quarter, so that a similar course may be followed. But, as Fig. 113 shows, the very property of a sharp minimum which enables us to use the direction finder for precise determinations renders it of little use for cutting out signals emanating from scattered sources: and the sources of disturbing atmospherics are always more or less scattered. Discrimination by direction is therefore of little avail against atmospherics.

(c) Provided that the unit atmospheric disturbance has very short duration compared with the unit signal, the Morse dot, its evil effects would be largely overcome if it rendered the receiver entirely inoperative. In any case it is better that the atmospheric should temporarily incapacitate the receiver than actuate it with great violence; for a violent noise in the telephones has a deafening effect on the operator which remains long after the noise itself has ceased.

This is the principle underlying the balanced rectifier arrangement of H. J. Round and the Marconi Co., which has proved to be of considerable utility. In Fig. 116, R_1, R_2 are two crystal or (better) thermionic rectifiers connected in parallel in opposite directions so that the signal current in the telephone is the difference of the two rectified currents. Separate potentiometer adjustments for the two rectifiers permit of the independent adjustment of the steady P.Ds. across them. Suppose, for simplicity, that they have the same I, V characteristic; and let P_1, P_2 be the respective representative points on the characteristic. If, as in the figure, P_1 is chosen for good and P_2 for bad rectification with weak signals, R_2 will have little effect in reducing the total rectified current with weak signals. With strong signals, however, R_1 and R_2 are almost equally effective rectifiers; so that the signal current in the telephone from a violent atmospheric is very much reduced by the presence of R_2.

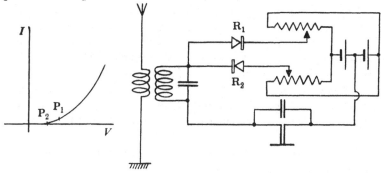

Fig. 116. Opposed rectifiers.

(d) Another class of anti-atmospheric device is illustrated in principle in Fig. 117. AC_1L_1 and AC_2L_2 are two antenna circuits, one tuned to the signal and the other slightly distuned. The atmospheric may be supposed to produce oscillations of equal amplitude in each circuit; the consequent rectified currents through R_1 and R_2 are equal and produce no sound in the telephone. The signal, on the other hand, would affect the tuned more than the distuned limb, and the rectified currents would not balance.

In the last arrangement, balancing is effected after rectification. The effects of an atmospheric may alternatively be balanced out at an earlier stage, viz. before rectification. It has

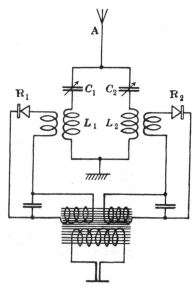

Fig. 117. Divided antenna circuit.

been observed that different types of antenna exhibit different values of the ratio

$$\frac{\text{sensitivity to lightly damped signals}}{\text{sensitivity to atmospherics}}.$$

It is therefore theoretically possible to combine the oscillations of two such antennae in such proportions that while the atmospheric effects cancel out the signal effects do not. A. H. Taylor has carried out many experiments on this principle. He has had considerable success with long-wave transatlantic signals, using a 24-turn loop, 30 feet by 77 feet, as one antenna, and 1,600 feet of insulated wire slightly submerged in sea water as the other*. Methods such as this involve the balancing of two high-frequency oscillations, and are necessarily difficult to realise because the balancing E.M.F's. must be not only equal in amplitude but also in exact anti-phase. They are characterised by the theoretical possibility of the receiver remaining sensitive to the desired signal at the very epoch during which the atmospherics are occurring. A great deal of work along these lines appears to be in progress.

(e) In theory, at least, a simple solution of the problem is discrimination by the more ordinary tuning processes *after* the application of a limiting device. The limiter is designed to prevent the atmospheric, however violent it may be, from having more than a certain maximum effect on the selective circuits. In the competition between signal and atmospheric, the latter is then unable to make up in strength what it lacks in syntony or lowness of decrement. The general principle of the method is illustrated

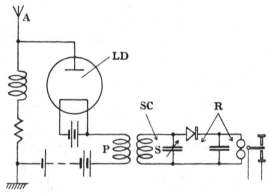

Fig. 118. Limiter-selector.

in Fig. 118. A is the antenna, which may be of any form, but must be highly damped so that it cannot be set into prolonged oscillation by the shock of an atmospheric. LD is a limiting device,

* *Proc. Inst. Radio Engrs.*, Dec. 1919.

preventing the production in the primary P of more than a certain high-frequency current however large the high-frequency P.D. across it. The secondary S is lightly coupled to P, and forms part of a selector circuit SC of low decrement. R is a relay or other form of trigger device actuated only when the oscillation in SC exceeds a certain threshold amplitude. Amplifiers may be inserted between A and LD, and between SC and R.

Since atmospherics certainly do not take the form of long lightly damped trains, it is theoretically possible in this way to prevent an atmospheric, however violent, from actuating R. While LD is exercising its limiting function, however, the desired signals as well as the atmospheric are cut off from R. But if during the Morse dot the proportion of time occupied by atmospheric E.M.Fs. in A is small—which is probably generally the case—R can be made to hold over to the marking position during the brief cessations of the signal due to atmospherics, and to register undamaged Morse signals in a Wheatstone receiver or other recording instrument.

This method does not appear yet to have been thoroughly put to the test. There are of course practical difficulties to be overcome. One is the production of a suitable limiting device LD, shown in the diagram as a thermionic diode. The ordinary simple diode or triode is a perfect current limiter, but only for signals which are very strong*; moreover oscillation in A must be prevented by suitable screening from affecting S otherwise than through LD, as for example by stray capacity effects. Nevertheless, these practical difficulties do not appear to be insuperable, and it is probable that some limiter-selector arrangement of this kind will prevent at least a large part of the interference at present experienced from atmospherics.

5. HIGH-FREQUENCY RESISTANCE OF CONDUCTORS

The well-known "skin effect" in conductors carrying alternating currents is especially prominent in wireless telegraphy because of the very high frequencies involved. The current varies in amplitude

---

* The author's "Kallirotron" is a possible sensitive limiter; *vide* "The Kallirotron: an aperiodic negative-resistance triode combination," *Radio Review*, April 1920. His "Oscillatory Valve Relay" (*Journ. Inst. Elect. Engrs.*, April 1920), also, is a sensitive trigger device which might conveniently replace SC and the insensitive rectifier with ordinary electro-mechanical relay shown in the figure.

and in phase at varying depths beneath the bounding surface of the conductor. Thus if an alternating current of frequency $\dfrac{10^7}{2\pi}$ periods per second ($\lambda = 189$ metres) flows in an indefinitely extended slab of conductor with one plane boundary—as in the water near the surface of the ocean, or in a wide thick copper plate— it may be shown that 95 per cent. of the whole current is contained within a sea depth of 1·4 metres or a copper depth of 1·4 millimetres. Accordingly the high-frequency currents, constituted by the moving ends of the electric lines depicted in Figs. 7, 8 and 9 as they sweep over the surface of the earth, do not penetrate far into the soil or the sea. A practical illustration of this fact is the rapid decrease in strength of the wireless signals received by a submarine as the depth of submergence is increased.

The same skin effect occurs in elongated conductors bounded in two dimensions, such as wires of circular or other cross-section. The current is not uniformly distributed over the cross-section, and in a thick conductor of low resistivity most of the interior may carry only a negligible current. The resistance of the conductor for high-frequency currents then greatly exceeds its resistance for steady or low-frequency currents. Mathematical investigations of the alternating current resistance of straight wires of circular section have been made by Rayleigh, Kelvin and others, and approximate formulae have been derived which are convenient for numerical calculations in certain practical cases arising in wireless telegraphy. Let

μ = permeability of material of wire;

ρ = resistivity of material of wire;

a — radius of cross-section of wire;

R_0 = resistance at zero frequency;

R_n = resistance at frequency n;

$$m \equiv \sqrt{\dfrac{\pi^2 \mu}{\rho}} \cdot a\sqrt{n}.$$

Then very approximately:

(i) $R_n = R_0 \cdot m \left(1 + \dfrac{1}{4m}\right)$ if $m > 2$

i.e. if $\dfrac{R_n}{R_0} > 2.25.$

(ii) $R_n = R_0 \left(1 + \dfrac{m^4}{3}\right)$ if $m < \cdot 8$

i.e. if $\dfrac{R_n}{R_0} < 1\cdot 14.$

The first formula is useful for stout wires of copper $\Big($for which

$\dfrac{\pi^2\mu}{\rho} = \cdot 076$ C.G.S.$\Big)$, and is applicable to sizes not smaller than

S.W.G. 24 at 10^6 periods per second,

and   S.W.G. 16 at 10^5 periods per second.

The second is useful for thin wires of copper, and especially for wires of the high-resistance alloys, such as Eureka or Constantan $\Big($for which $\sqrt{\dfrac{\pi^2\mu}{\rho}} = \cdot 014$ C.G.S.$\Big)$. It is applicable in copper to sizes not larger than

S.W.G. 35 at 10^6 periods per second,

and   S.W.G. 23 at 10^5 periods per second;

and in Eureka or Constantan to sizes not larger than

S.W.G. 19 at 10^6 periods per second,

and   S.W.G. 9 at 10^5 periods per second.

Fine Eureka or Constantan wires are often used in the laboratory for constructing high-frequency resistances of known value. From formula (ii) it may be calculated that the high-frequency resistance does not differ by more than 2 per cent. from the steady current resistance if the size of the wire does not exceed

$a = \cdot 035$ cm, say S.W.G. 22, at 10^6 periods per second,

and $a = \cdot 112$ cm, say S.W.G. 14, at 10^5 periods per second.

That large solid copper wires are uneconomical for carrying high-frequency currents is well exhibited in Fig. 119*, where the conductance expressed as metres per ohm is plotted against the cross-section in square millimetres, for steady current and for the two frequencies 10^6 and 10^5 periods per second ($\lambda = 300$ and 3,000 metres).

In the straight circular wire, the increase of resistance at high frequency is due to the non-uniform distribution of current along

* Calculated from tables given by J. Zenneck, *Lehrbuch der dr. Teleg.*

a radius of the cross-section. In the straight wire, however, the distribution is the same along every radius. When the wire is coiled into a close helix to form an inductance coil, this symmetry no longer obtains, and the current is denser towards the inside of the helix. The resistance is thus still further increased, and in practice the helical resistance may be several times the straight resistance.

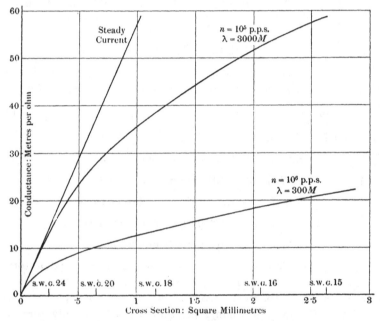

Fig. 119. The skin effect in copper wires.

In order to constrain the current to distribute itself uniformly, and so to obtain a low resistance despite the high frequency, it is a common practice to construct stout conductors in the form of a number of fine strands separately insulated and cabled together in such a manner that each strand is rotated and passed from inside to outside of the cable in the same way as every other strand. In stranded coils of this sort, high quality of material and workmanship is required. Leakage or dielectric loss in the insulating covering (e.g. enamel) of the strands, or imperfections in the spiralling of the separate strands, may easily lead to a costly stranded coil having a resistance higher than that of a cheaper coil of the same size wound of solid wire. In any case, the ratio

$\dfrac{R_n}{R_0}$ for a single strand within the coiled cable is greater than its resistance when isolated*.

6. The present trend of development

Wireless telegraphy, and even more telephony, is passing through a stage of very rapid development, largely owing to the introduction of the high-vacuum triode. The enormous increase in sensitiveness of receiving apparatus has made possible a great extension in range of communication, or a great reduction in transmitter power, or a great reduction in size of receiving antenna, or a combination of such changes. To these might be added increase of selectivity, and automatic reception with increase of speed. But although these receiver improvements have made it possible to construct remarkably small and portable short-range stations (as for military purposes), yet in long-range work there has been no tendency to reduce transmitter power. The following extract from paragraph 14 of the Report of the "Imperial Wireless Telegraphy Committee, 1919–1920" clearly states the historical position in this respect.

...In arriving at our conclusions we have been largely guided by the history of long-distance transmission by wireless telegraphy. This begins in the year 1901, when by means of newly-discovered methods Mr Marconi attempted to establish communication across the Atlantic, a distance of more than 2000 miles, by a spark station of about 25 kilowatts. Experience gained on short distances had suggested that this power would be ample. But factors which become prominent only at great ranges presented unexpected difficulties, and the power of the transmitting station was increased step by step in successive attempts until in 1906 Marconi was using a power of about 75 kilowatts. In spite of the confidence of the pioneer workers, the telegraphic service obtained was liable to frequent interruptions through natural causes, and still greater power stations were projected. This was necessary although the sensitiveness of the methods of reception of signals had been enormously enhanced since the original estimate of 25 kilowatts was made. In 1906 the Poulsen arc was introduced into practical wireless telegraphy, and completely new methods were advocated by its promoters. From their experience gained in trials across the North Sea between Denmark and England the Poulsen group anticipated complete success across the Atlantic with about 15 kilowatts. This estimate once more proved too low. Meanwhile the early pioneers and other inventors in various countries pushed the spark methods to higher

* For a discussion of the high-frequency resistance of solid and stranded coils, see G. W. O. Howe, "High-frequency resistance of wires and coils," *Journ. Inst. Elect. Engrs.*, Feb. 1920.

and higher power, till in 1912 spark stations of more than 200 kilowatts existed for transmission across distances of about 2300 miles. Even yet interruptions were frequent. Valuable observations were made about this period by engineers in the service of the American and other Governments on both arc and spark stations, and it appeared that with the very best apparatus the telegraphic service over these ranges, or even smaller ranges in the tropics, could not be continuously maintained.

During the years 1913 to 1919 the methods of reception have, as has already been mentioned, increased vastly in sensitiveness. It has therefore been thought that surely now the large-power continuous-wave stations erected in America and several countries of Europe just before and during the war will be able to maintain uninterrupted telegraphic service across distances of about 3500 miles. Our enquiries have convinced us that arc stations employing 250 kilowatts are not sufficient to ensure an uninterrupted twenty-four hours' service over such distances throughout the year even in the temperate zone, in spite of many elaborate methods for eliminating natural disturbances having been developed and applied. The demand is still for more power at the transmission stations. At present the largest transmitting station working across the Atlantic is the German station at Nauen, which is rated at 400 kilowatts, and which aims at communication across 4000 miles, but it is understood that the German engineers have recently decided upon a great increase in the power of this station.

. . . .

We are convinced that consideration for commercial purposes of lines like those from England to India, from India to Australia, and from Australia to Canada, involves far-reaching speculation beyond existing experience. It is noteworthy, too, on the one hand, that mathematical theory tends to emphasise the need for great caution in embarking upon such speculation, and on the other, in practice, that when the leading American experts, as has been previously stated, desired in 1918 to ensure absolute regularity of service between France and the United States, they started to erect an arc of no less than 1000 kilowatts*, and this for a distance not nearly that between England and India, and much less subject to atmospheric disturbance than India.

The explanation of the demand for ever greater power at the transmitter is mainly to be found in interference from atmospherics, discussed in Section 4 of this chapter. If atmospherics were wholly absent, the receiver would be subject to interference only from the stray signals of other wireless stations. Technically, if not politically, it would then be possible to decrease the power in all transmitting stations *pari passu* with increases of sensitivity in all receiving stations, without change in the efficiency of the service. In the presence of atmospherics, however, if the full advantage is to be obtained, increase in sensitivity of receiving

* At Croix d'Hins, near Bordeaux; now being completed by the French Government.

apparatus must be accompanied by corresponding advances in anti-atmospheric devices. In recent developments our powers of selection have lagged far behind the advances in our powers of detection.

At the receiver, then, the direction of development is likely to be improvement in means of combating atmospheric interference. Success here will be followed by the general adoption of automatic reception at high speeds. As far as can be foreseen, circuits of low decrement will continue to be necessary for conferring selectivity as between transmitting stations. The attainable speeds are therefore likely to be limited ultimately by the approach of the duration of the Morse dot to the parameter $\frac{1}{n\delta}$ of the oscillatory circuit*.

At the transmitter, ever greater power will be demanded until the fight against atmospherics has been won. When the engineer succeeds in evading rather than shouting down the interference, a limit will have been set; and π times the radius of a well ordered spherical world would be the *maximum* range of the largest stations on it.

* See Chapter III, and especially footnote on p. 20.

INDEX

Printed in the United States
By Bookmasters